OECD Studies on Water

Water Risk Hotspots for Agriculture

This work is published under the responsibility of the Secretary-General of the OECD. The opinions expressed and arguments employed herein do not necessarily reflect the official views of OECD member countries.

This document, as well as any data and any map included herein, are without prejudice to the status of or sovereignty over any territory, to the delimitation of international frontiers and boundaries and to the name of any territory, city or area.

Please cite this publication as:
OECD (2017), *Water Risk Hotspots for Agriculture*, OECD Studies on Water, OECD Publishing, Paris.
http://dx.doi.org/10.1787/9789264279551-en

ISBN 978-1-78040-936-8 (print)
ISBN 978-92-64-27955-1 (PDF)

Series: OECD Studies on Water
ISSN 2224-5073 (print)
ISSN 2224-5081 (online)

The statistical data for Israel are supplied by and under the responsibility of the relevant Israeli authorities. The use of such data by the OECD is without prejudice to the status of the Golan Heights, East Jerusalem and Israeli settlements in the West Bank under the terms of international law.

Photo credits: Cover © tomasworks/Shutterstock.com

Corrigenda to OECD publications may be found on line at: *www.oecd.org/about/publishing/corrigenda.htm*.

© OECD 2017

You can copy, download or print OECD content for your own use, and you can include excerpts from OECD publications, databases and multimedia products in your own documents, presentations, blogs, websites and teaching materials, provided that suitable acknowledgement of OECD as source and copyright owner is given. All requests for public or commercial use and translation rights should be submitted to *rights@oecd.org*. Requests for permission to photocopy portions of this material for public or commercial use shall be addressed directly to the Copyright Clearance Center (CCC) at *info@copyright.com* or the Centre français d'exploitation du droit de copie (CFC) at *contact@cfcopies.com*.

Foreword

Increasing evidence suggests that water risks threaten future agriculture production in many regions. Factors, such as the multiplication of extreme water events, sea level rise—both projected to accelerate with climate change—water quality deterioration, groundwater depletion, and intensifying cross-sector competition for water supplies, combine to create "a perfect storm" for agriculture in many regions, which are often poorly prepared to respond. If these water risks are particularly intense in specific agricultural regions, their impacts can expand to national and international levels, with consequences on markets and food security.

This report recommends using a hotspot approach to respond to these future water risks. Targeted actions in identified regions at risk can help increase efficiency and effectiveness of the sector's response. Farmers, agro-food companies and governments all have a role to play in risk mitigation efforts. Governments should reinforce their water management actions in the regions facing water risks, but they should also strengthen markets and trade, and encourage international collaboration to limit the diffusion of impacts from these risks.

The study builds on past OECD work on water and agriculture; especially recent reports on climate change, water and agriculture, managing groundwater use in agriculture, and mitigating drought and floods in agriculture. It also relies on three new analyses: a global assessment of water risk hotspots for agriculture production, a "water stress test" simulation of production and market impacts in three hotspot regions and a micro-economic model outlining how farmers and agro-food may respond to different water risks.

Acknowledgements

This report was written by Guillaume Gruère (OECD), with contributions from multiple partners and colleagues. Discussions on extreme weather events in Chapter 1 and Chapter 3 are based on a note written by Jean-Joseph Cadilhon (OECD). Chapter 2's discussion on defining water risks relies on a consultant report written by Paul Reig and Tien Shiao of the World Resource Institute (WRI). Three case studies support the analysis in Chapters 2 and 3; a report written by Heather Cooley, Michael Cohen, and Rapichan Phurisamban of the Pacific Institute on the Southwest United States (published separately); a literature review on Northeast China written by Laura Munro and Juliane Jansen (OECD); and a literature review on Northwest India written by Alice Lucken (Paris Institute of Political Studies). The simulations of water risks in Chapter 3 were led by Laetitia Leroy (consultant) with support from Daniel Mason D'Croz (IFPRI) and IFPRI's IMPACT team. Evelyn Farris (Graduate Institute of Geneva) and Alice Lucken provided additional research support on Chapters 2 and 3. Feedback from participants of the workshop "Managing water risks for agriculture: a discussion with the private sector", organised by the OECD in co-operation with the Dutch Ministry of Economic Affairs in November 2016, and summarised by Assia Elgouacem (OECD), was used to develop Chapter 4's discussion on private sector involvement.

More generally, the report benefitted from comments and suggestions from OECD delegations, from OECD colleagues Robert Akam, Rachel Bae, Charles Baubion, Jean-Joseph Cadilhon, Carmel Cahill, Matthew Griffiths, Julien Hardelin, Ada Ignaciuk, Franck Jésus, Hannah Leckie, Xavier Leflaive, Laura Munro, Michèle Patterson, and Janine Treves, and from Daniel Mason D'Croz and Tim Sulser of IFPRI. The report was copy-edited by Caitlin Connelly and Natasha Cline-Thomas. Theresa Poincet helped manage the administrative process, Noura Takrouri-Jolly helped format the figures, and Stéphanie Lincourt formatted the report for publication.

Table of contents

Tables

Figures

Boxes

Executive summary

Agriculture is expected to face increasing risks that stem from water shortages, floods and the degradation of water quality, all of which will have an impact on production, markets, trade and food security. Available freshwater is projected to be constrained by a growing demand for water beyond the agricultural sector and increased variability of precipitation due to climate change. Extreme weather events may become more frequent. Water quality will likely deteriorate in many regions. These changes are expected to strongly impact agriculture, a highly water-dependent sector, and, in turn, affect the production and productivity of rainfed and irrigated crop and livestock activities.

Targeted policy actions that address the specific needs of localised agriculture productive regions subject to acute water risks, i.e. "hotspots", could increase the efficiency and effectiveness of policies that seek to mitigate future water risks. Agriculture activities and water resources vary geographically, and consequently, the sector's vulnerability to and impacts on water risks differ by region. Targeted mitigation of water risks via hotspots can produce more cost efficient results, help prevent the diffusion of risks, and focus attention and efforts on specific regions. This hotspot approach can be applied at virtually any scale for individual or multiple water risks, and can account for all dimensions of water risks in a limited number of spots. Such targeted policy action can also mitigate the diffusion of water risks via indirect effects such as supporting the customisation of agriculture climate adaptation plans at the local level.

This report outlines the hotspot approach, presents an application at the global scale, and puts forth a policy action plan to mitigate the future impact of water risks on agriculture. This approach differs from traditional policy recommendations in that it aims to focus and prioritise attention and investment on risky areas, and to limit the impact on agriculture production, markets and food security.

Three conditions are necessary to identify hotspots for future risk assessments. First, there should be well-defined risk peaks, i.e. well-defined regions that are subject to high risks compared to surrounding areas. Second, the limits of the hotspot areas need to be selected to ensure a balance between efficiency and comprehensiveness objectives. Third, any assessment requires sufficiently robust information to avoid costly mistakes, such as the possibility of hotspot omission or erroneous inclusion.

An assessment of future water risk hotspots for global agriculture production based on a comprehensive assessment of the water risk literature, combined with baseline projections on agriculture production in 2024 and 2050, identifies three countries: the People's Republic of China (hereafter "China"), India and the United States. These countries lead global production of major commodities, but are also considered to be the most exposed to future water risks of the 142 countries and economies covered in this study. A number of other countries and regions are expected to face high agriculture water risks, but the water risks that agriculture production will face in these three countries could have global consequences.

Within each of these countries, agriculture water risks are especially prevalent in the regions of Northeast China, Southwest United States and Northwest India. Northeast China — a key region for cereal production — is a drying region, with water quality issues, groundwater depletion, and growing and competing water demands from industry and urban growth. The semi-arid region of Northwest India is known as India's breadbasket, with large wheat and rice production, but this production is supported by intensive groundwater irrigation with worrying current and projected consequences for the water table and for the region's water quality. The Southwest United States region — a very productive and diverse agriculture region that produces dairy, livestock, fruits and vegetables — is also facing drier and warmer conditions, severe groundwater depletion, and projected competition for water demand from rapid future population growth.

When agriculture water risks become reality in locations identified as hotspots, there can be three levels of impact: (1) a local fall in production; (2) an adverse effect on agriculture markets and trading partners; and (3) increased tensions over food security and socio-economic issues among a larger set of countries.

The evidence collected shows that these impacts can be significant:

- Water risks can affect **agriculture production** significantly in hotspot regions. For instance, projections show that in the absence of policy action, there will be significant negative impacts on agriculture production in Northeast China, Northwest India and the Southwest United States due to the shortage of water. Activities that generate low economic value per water use will be the first affected by water risks.
- Water risk hotspots can have **domestic and global market repercussions**. A simulation of projected impacts of gradual increases in surface and groundwater irrigation stresses and droughts in the three regions, with no water shock elsewhere, showed reduced global production and increased prices of major field crops, in particular maize, wheat and cotton, but also fruits and vegetables. The simulation suggests that national production in these three countries could fall by a few to a dozen percentage points, affecting their trade balance as well as trade balances with partner countries. Some of these effects could intensify with climate change.
- Acute agriculture water risks can also have **broader food security and socio-economic consequences**. High agriculture water risks partially explain the multiplication of foreign land purchases, often conducted by water scarce countries into relatively better water-endowed countries. These risks can also result in social tensions, fuelling conflicts that can become regional, and they can drive migration both domestically and internationally.

Farmers, agro-food companies, and governments all play a role in responding to water risks in hotspot locations. Farmers are more likely to respond to high water risks they are expected to confront in the medium term (e.g. drought) than to water risks to which they contribute (e.g. groundwater pumping or pollution). Large private companies may be more willing than small companies to address water risks they create themselves, through their purchase of products from farms using unsustainable agriculture practices. Governments have a role in encouraging farmers and companies' actions, strengthening resilience and reducing future damages in hotspot locations.

A three-tier policy action plan is proposed to address water risk hotspots.

1. **Prioritise action in hotspot regions**: As a priority, water-risk country governments should focus their attention and efforts on hotspot regions, in particular through the introduction of targeted agriculture and water instruments to improve data collection, disseminate best agriculture practices, encourage technological innovation, and improve governance structures. In addition, they should consider customising water and agriculture policies to hotspot regions by reinforcing local efforts or adapting economic instruments and by co-ordinating their efforts with private agro-food companies and other water using sectors.
2. **Strengthen market and trade relationships:** To limit the effects of water risk hotspots on agriculture markets domestically and globally, governments should work with trade partners to reinforce the links between domestic and international markets to support integration and competition, and to reduce trade barriers to limit domestic market impacts in hotspot countries.
3. **International collaboration:** International collaboration should be developed to bolster resilience to future water risks and information exchange in order to reduce diffusion of indirect impacts from hotspot countries and regions, and to help countries not facing water risks prepare themselves against unexpected indirect effects associated with water risks occuring in other countries.

Chapter 1

Addressing water risks in agriculture

Trends and projections suggest that agriculture will face increasing water risks in many regions which could affect agriculture production and markets as well as international trade and food security. This chapter presents evidence on future water risks for agriculture and introduces the hotspot approach as a means to assess and respond to these risks.

Key messages

Due to a combination of climatic constraints, current water uses, and rising competition for water, agriculture in many regions is projected to face multiple water risks that could negatively affect local, regional and global food production and food security. Water shortages, excessive water and water quality deterioration are projected to increase in some regions and will have an impact on agriculture production.

Agriculture both contributes to and faces such risks. Water is essential to agriculture production, yet in many countries this sector is a large and inefficient user of water. It is also a large polluter of surface and groundwater.

The use of a targeted approach, e.g. focusing actions on particular areas, to future agriculture water risks could help better cope with these challenges, while saving public resources. As outlined in this chapter, the report assesses and proposes policy responses to mitigate the effects of acute future water risks in localised productive agriculture regions, i.e. “hotspots”, for agro-food systems and markets.

1.1. Past and recent trends demonstrate the growing importance of water risks

Recent studies show increasingly high trends in water risks.[1] Global assessments report increasing risks of shortages and flooding (Sadoff et al., 2015). The global rise in overall water demand and increased groundwater use (top of Figure 1.1) have contributed to rising tensions between the use of freshwater supply and dealing with regional water scarcity. Concurrently, average sea levels have continued to rise and extreme flooding events are more frequent (bottom of Figure 1.1), trends which are partially attributed to climate change (Jimenez-Cisneros et al., 2014).

Figure 1.1. Worrying water trends

Top left: Global water demand (1960-2000), Top right: Groundwater abstraction and depletion (1980-2000)

Bottom left: sea level rise (1993-2015), Bottom right: frequency of extreme flooding events (1983-2012)

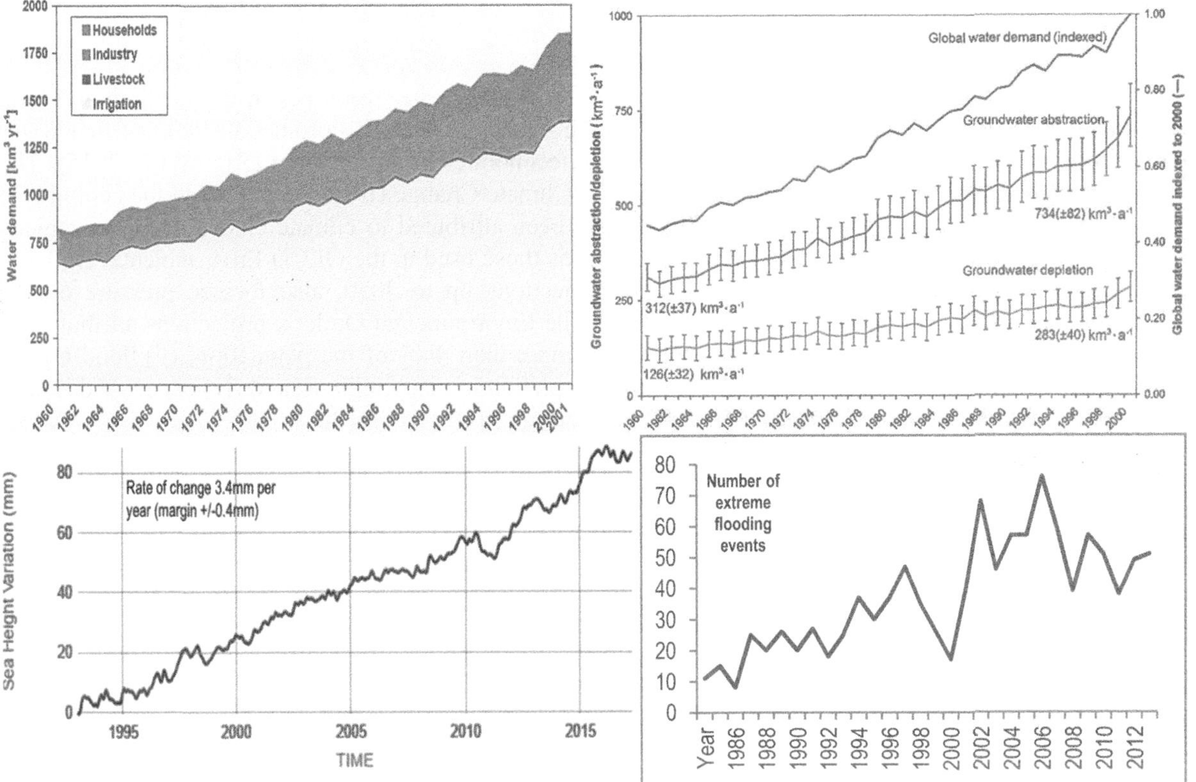

Source: Derived from Wada et al. (2011); Wada et al. (2010); NASA Goddard Space Flight Center (2017), https://climate.nasa.gov;; "Global Active Archive of Large Flood Events", Dartmouth Flood Observatory, University of Colorado, http://floodobservatory.colorado.edu/Archives/index.html.

The annual number of disasters triggered by weather-related natural hazards more than doubled from the 1980s to the first decade of the 2000s (FAO, 2015). Droughts and related heatwaves are more frequent in many regions (OECD, 2014a). In 2015, droughts were particularly severe in southern Brazil, southwestern United States and northern Chile, and severe droughts started in southern Africa and Thailand. Trend observations from the Dartmouth Flood Observatory suggest that extreme floods have also become more frequent, with their impact also increasing over time.

Extreme weather events previously considered as seasonal occurrences such as storms and cyclones, have become more frequent and severe in many regions of the world (Herring et al., 2015). For instance, the number of storms in the Eastern North Atlantic increased by 200% during between December and March 2014; cyclones in the Bay of Bengal have become significantly more frequent in the northern part of the Bay and stronger than normal after the regular monsoon period. Extreme rain events like the one observed in 2014 in the French Cevennes are three times more likely today. Above average yearly maximum rainfall are being seen in the Jakarta area where they are twice as likely to occur today as in the past.

The agriculture sector is bearing a large part of the losses and damages associated with these water-related extreme events (OECD, 2016). The Food and Agriculture Organization (FAO) has estimated that 30% of all damages caused by 140 weather-related natural disasters between 2003 and 2013 in developing countries were borne by agriculture (FAO, 2015). Agriculture accounts for 84% of the economic impact of droughts (Ibid.). Droughts, cyclones, floods and cold waves have led to agriculture crop and livestock production losses worth USD 80 billion. In OECD countries, droughts have been particularly damaging for agriculture, leading to reductions in crop yields and farm revenues, or to large increases in crop insurance compensation (OECD, 2016). More generally, Lesk et al. (2016) estimate that severe droughts and extreme heat events between 2000 and 2007 were responsible for crop losses equivalent to 6% of global cereal production.

1.2. Water risk projections depict a bleak future for agriculture in many regions

There is increasing concern about the availability of usable water resources in the medium and long term. In its 2015 ten-year risk-landscaping exercise, the World Economic Forum (WEF) identified water crises as potentially having the single greatest impact on economies in the medium term (WEF, 2015). The fifth assessment report of the Inter-government Panel on Climate Change (IPCC) emphasised the central role of water risks, noting that significant water effects have been attributed to climate change in many regions (Pachauri et al., 2014). Business-as-usual scenarios, such as those used in the OECD Environmental Outlook, combining climate impacts on supply and demand projections up to 2050, also foresee pressure on the quantity of fresh water and deteriorating water quality. The Environmental Outlook projections attributed its highest risk ratings to (i) quantitative water stresses for more than 40% of the population, (ii) flood risk to nearly 20% of the world's population, (iii) groundwater depletion in many regions, and (iv) expected decrease in the quality of surface water, especially in non-OECD countries (OECD, 2012a).

Regional and country analyses confirm these assessments (OECD, 2014a). On the quantitative side, existing evidence suggests that the water cycle will accelerate with more frequent episodes of strong precipitations in high latitudes and fewer in lower latitudes (Buckle and Mactavish, 2013; OECD, 2014a).[2] For example, Southern Europe is expected to face a significant reduction in the availability of surface water (Forzieri et al., 2014), coupled with increased seasonal variations in precipitation (more intense in winter), resulting in negative groundwater recharge overall (MEDDE, 2012). Australia will be increasingly vulnerable to rises in sea level (OECD, 2014a), while flooding costs in Europe could increase five-fold by 2050 (Jongman et al., 2014). By 2030, baseline scenarios find flood risks doubling in urban or rural areas, with the greatest risks in Asia (Sadoff et al., 2015). Long-term water management decisions upstream may also result in exacerbated regional water risks downstream in other countries, with significant economic, environmental and social consequences (Orr et al., 2012).

Water quality is projected to deteriorate in multiple regions due to human activity and climatic factors. Increased discharge of nitrogen and phosphorus from agriculture and other sectors resulting in eutrophication are projected particularly in parts of Sub-Saharan Africa, India and Southeast Asia (IFPRI and Veolia, 2015). These changes will be exacerbated under a drier climate.[3] Climate change-induced sea level rises will increase the risk of saline intrusion in coastal aquifers (Jiménez Cisneros et al., 2014), affecting the quality of groundwater in multiple regions, including Japan, Mexico and the Netherlands (OECD, 2015). Dryland salinity is expected to expand over the next few decades in Australia because of changes in precipitation patterns (Hart et al., 2003).

These water risks are expected to strongly impact agriculture, a highly water-dependent sector (UNEP, 2016). Agriculture accounts for approximately 70% of the world's water withdrawals and 85% of global freshwater consumption (Pfister and Bayer, 2013). Roughly 80% of cultivated land across the globe is exclusively rainfed, accounting for 60% of the world's crop production (FAO, 2009). The remaining

20% obtains water from irrigation systems to meet all or some portion of crop-water demand, contributing to 40% of global crop production.

All types of agriculture activities could be affected (Figure 1.2). Rainfed agriculture will face changes in precipitation patterns, potentially increasing frequency of extreme events due to climate change (Box 1.1). Irrigated agriculture will face similar types of variability in water supply and demand, with an increase in water demand and competition from other sectors, and water quality challenges threatening its viability (HLPE, 2015).[4] Groundwater depletion in some semi-arid areas may affect the productivity potential of major cropping systems and their resilience to climate change (OECD, 2015). It is expected that declines in water quality, due in part to irrigation-induced salinity and soil erosion, will affect freshwater availability for agriculture (OECD, 2012b). Rising seawater levels will affect coastal croplands and deltas (ADB, 2011), and droughts and floods will affect livestock productivity (OECD, 2014a).

Figure 1.2. Agriculture production is projected to face a combination of water risks

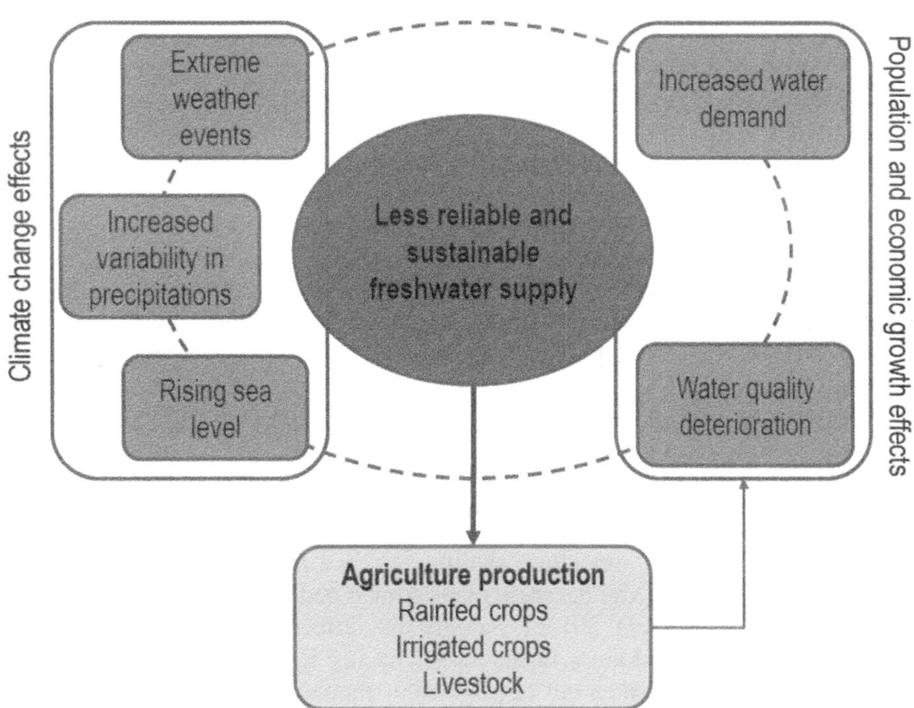

Note: Arrows represent the main impacts of water risks on agriculture and of agriculture on water risks.

Source: Author's own work.

Box 1.1 Water risks for agriculture under climate change

Climate change will have multiple impacts on the water cycle for agriculture. Three variables are critical to agriculture: higher temperatures, future precipitation patterns and their distribution throughout the year, and the incidence of extreme weather events. The main consequences of changes in water resources for agriculture production include:

- Increased crop evapotranspiration due to rising temperatures (up to a maximum heat threshold), which may increase crop demand for water.
- Increased water shortages, particularly during the spring and summer months with increased water requirements for irrigation. This will be especially difficult in areas already suffering from water stress.
- Reduced snowpack, affecting the seasonal flow of surface water for irrigators.
- Expansion of crop irrigation calendars raises irrigation requirements.
- Increased risk of flooding due to the expected concentration of rainfall during the winter months.
- Reduced water quality due to higher water temperatures and lower levels of water runoff in some regions, imposing further stress in irrigated areas. Other factors include increased sediment, excessive nutrient, and pollutant loadings from heavy rainfall.
- Increase in sea levels affecting agriculture production in low-lying coastal areas.

Source: OECD (2014a), Iglesias and Garrote (2015), Fischer et al. (2007).

As a key agriculture input, the availability of usable water will have an impact on production at the local level as well as on markets and consumers on a much wider scale. Multiple studies have shown that irrigated crop yields are significantly higher than those from rainfed agriculture (Hertel and Liu, 2016; WWAP, 2012). It also generates more value, for instance in the United States, in 2012, the average value of farm products by irrigated farmers was about 3.9 times the average value for non-irrigated farms (Schaible and Aillery, 2016). This could have both local and global repercussions, affecting local production, which relies on irrigation, and broader agriculture markets and consumers. In particular, water risks related to climate change will contribute to exert pressure on agriculture markets by increasing commodity prices (Ignaciuk and Mason D'Croz, 2014; Ignaciuk et al., 2015). For instance, Sadoff et al. (2015) estimated that ensuring "full water security" for irrigated agriculture[5] would decrease the probability of global wheat production falling below 650 million tonnes per year from 83% to 38%, and the probability that the price of rice could exceed USD 400 per tonne from 21% to 0.7%. Overall, water security for irrigation could generate welfare benefits of USD 94 billion (Ibid.).

Water quality could also impact agriculture significantly in a number of ways (OECD/Ringler, 2011; OECD, 2014a). Reduced water quality can decrease plant growth and increase livestock contamination, thus affecting agriculture productivity (OECD, 2013b). Salinity may affect up to 20% of irrigated areas globally and threaten nearly half of all irrigated areas in the long term (Le Kama and Tomini, 2013). One and half million hectares of agriculture land is taken out of production annually as a result of land salinity with total costs for producers potentially exceeding USD 11 billion per year (Schoengold and Zilberman, 2007). Plants can absorb toxic chemicals from contaminated water and pass them on to humans and animals who consume them (OECD, 2013b). Toxic contamination of water bodies may poison livestock and generate losses in the meat processing and milk industries. For example, 80% of the milk supply in Hawaii was disrupted in 1982 following water contamination by the insecticide heptachlor, generating a loss of USD 8.5 million for the milk industry (Pimentel, 2005).

These water risks, in isolation or in combination, could have much wider food security implications (Hanjra and Qureshi, 2010; Ringler et al., 2010), "threatening the sustainability of livelihoods dependent on water and agriculture" in key production regions (FAO and WWC, 2015). Stresses on groundwater resources in highly populated irrigated areas could result in a significant impact on food security for millions (Famiglietti, 2014). Food-insecure regions, especially in parts of Sub-Saharan Africa, which depend on rainfed agriculture, could further suffer from repeated droughts (Ringler et al., 2010).[6] Fisheries in the Mekong River basin will suffer from any shortage of water associated with upstream dam-induced water stresses, with millions of people at risk of losing critical protein and income sources (Orr et al., 2012). Irrigation will likely have to play a more important role in agriculture in some of these regions, but in a context of increased competition for water both within and outside the sector (Ibid.).

Acute water risks in agriculture regions will also impact the environment. Ecosystems in rural areas could be directly affected by droughts (OECD, 2016). Associated with heat, droughts facilitate wild fires that ravage entire forest systems. Water quality will decrease in regions with water shortages due to the increased concentration of pollutants further impacting water dependent species (OECD, 2012a). Extreme weather events may also damage the environment and limit its resilience to future shocks.

1.3. How can policy respond to emerging water risks?

While agriculture is expected to be impacted by future water stresses, it can also play a significant role in mitigating these risks. The agriculture sector is a significant emitter of greenhouse gas and therefore has a role to play in addressing human-induced climate change (MacLeod et al., 2015). Agriculture is also the largest freshwater consuming sector intrinsically linked to improved water resource management, which can happen via a more efficient use of water in addition to reducing the sector's negative impact on water quality (OECD, 2014a). Irrigation, for example, has been supporting agriculture productivity, but pollution from large irrigation schemes has affected 34 million hectares globally to date (Mateo-Sagasta and Burke, 2011). In most cases the effect of irrigation on groundwater recharge will be positive; however, groundwater quality can be impacted by diffuse pollution from agriculture practices such as fertiliser and manure spreading (Böhlke, 2002). Better agriculture and water management policies can help mitigate these impacts as well as those stemming from droughts and floods (OECD, 2016).

These interlinked challenges are often not adequately addressed by national policies due to their complexity and the importance of local differences. A targeted or "hotspot" approach could be an effective and cost efficient way to address future agriculture water risks. It would require the identification of specific areas where future agricultural water risks are expected to be the greatest within a particular geographic or administrative area and the use of targeted actions to limit potential negative impacts.

This report aims to assess and propose policy responses to mitigate the effects of acute future water risks in localised productive agriculture regions (hotspots) for agro-food systems and markets. The proposed hotspot approach is defined to be used at any geographical or administrative level. At the same time, the report specifically considers hotspots that can have cross-border implications, analysing direct future risks to agriculture production in water stressed countries, but also indirect risks via market effects driven by deteriorating water conditions in other countries.

This report builds on OECD work on climate change, water, and agriculture (OECD, 2013a and 2014a), water quality (OECD, 2012b), the management of agriculture risks (OECD, 2009 and 2011), effective targeting of agriculture policies (OECD, 2007), and water security (OECD, 2013b).[7] It adds to the existing body of work in four ways.

The report addresses three questions:

1. *Where are the future water risk hotspots for agriculture?* This report defines the hotspot approach and then applies this method to identify some of the most pressing geographic and commodity-specific water risk hotspots for agriculture production, agro-food systems and markets, globally, using available data, and acknowledging uncertainties in models and projections, trade-offs in scale of analyses, and types of risks.

2. *What are the implications of water risk hotspots for agriculture production and markets, and more widely for food security?* This report offers an analysis of the expected impacts on agriculture markets related to water risk hotspots in hotspot and non-hotspot areas, and the broader consequences they may have on food security.

3. *How can public policy mitigate these risks?* The role, effectiveness and efficiency of current and future public policies to manage risks is analysed in hotspot and non-hotspot areas, taking into account endogenous farm-level and market-driven adaptation actions, as well as forward-looking actions by private agro-food sector companies.

Chapter 2 discusses the relevance and methods to use the hotspot approach and applies the approach by conducting an assessment of global future water risk hotspots for agriculture production based on existing literature and agriculture baseline projections. Chapter 3 analyses the implications of water risk hotspots for the agro-food sector and broader food security. Chapter 4 proposes a policy action plan to mitigate critical future water risks and their propagation at the local, market and broader level.

Notes

1. Recent research has used refined methods to evaluate water risks at the global, regional or local level. In particular, better data and more sophisticated analyses have measured groundwater depletion rates (e.g. Famiglietti, 2014; Gleeson et al., 2012), identified regions most subject to floods or droughts damages (Sadoff et al., 2015), and evaluated the irrigation area, cities and population facing water scarcity (e.g. Hanasaki et al., 2008; Brauman et al., 2015; Mekonnen and Hoekstra, 2016).

2. The IPCC projects that climate change will "reduce renewable surface water and groundwater resources in most dry subtropical regions (robust evidence, high agreement), intensifying competition for water among sectors (limited evidence, medium agreement)" (Pachauri et al., 2014; Jimenez Cisneros et al., 2014).

3. Climate change is projected to increase water temperatures and precipitation intensity, and induce longer periods of low flows which will "exacerbate many forms of water pollution, including sediments, nutrients, dissolved organic carbon, pathogens, pesticides, salt and thermal pollution" (Bates et al., 2008).

4. Surface water or groundwater may be subject to risks of depleting quality due in particular to inadequate wastewater treatment or pollution activities making it unusable.

5. Defined here as the absence of water security constraints (quality, quantity) for agriculture.

6. For instance, Pauw et al. (2010) show that droughts and floods result in losses of 1.7% GDP due to their effects on agriculture. Small-scale farmers in lowlands are the most affected by floods. Agriculture losses can create food shortages and price increases with direct effect on urban populations.

7. In a broader policy setting, OECD (2010) provided strategic management approaches to future shocks, including extreme events and natural catastrophes. The report served as a foundation for the OECD recommendation on critical risks (OECD, 2014b), outlining five necessary axes of governance: supporting a comprehensive approach to critical risks, ensuring preparedness, raising awareness, and building adaptive capacity in crisis management, all to be done in a transparent and accountable manner.

References

ADB (Asian Development Bank) (2011), "Food security and climate change in the Pacific: Rethinking the options", Pacific Studies series, ADB, Manila, the Philippines.

Bates, B.C. et al. (eds.) (2008), "Climate Change and Water", Technical Paper of the Intergovernmental Panel on Climate Change, IPCC Secretariat, Geneva.

Böhlke, J. K. (2002), "Groundwater recharge and agricultural contamination", *Hydrogeology Journal*, Vol.10, pp.153-179.

Brauman, K. A. et al. (2015), "Water depletion : An improved metric for incorporating seasonal and dry-year water scarcity into water risk assessments", *Elementa: Science of the Anthropocene*, Vol. 4, pp. 1–12. doi:10.12952/journal.elementa.000083

Buckle, S. and F. Mactavish (2013), "The changing water cycle", Grantham Briefing Note 5, University College of London, London.

Ciscar, J.C. et al. (2014), "Climate Impacts in Europe", The JRC PESETA II Project, Joint Research Centre (JRC) Scientific and Policy Reports, EUR 26586EN, European Commission, Seville, Spain.

Famiglietti, J. (2014), "The global groundwater crisis", *Nature Climate Change*, Vol. 4, pp. 431-438.

FAO (2015), The Impact of Disasters on Agriculture and Food Security, FAO, Rome. http://www.fao.org/resilience/resources/resources-detail/en/c/346258/

FAO (2009), "Resource outlook to 2050: Water use in irrigation and pressures on water resources", FAO, Rome.

FAO (2007), "Coping with water scarcity, Challenge of the 21st Century", FAO, Rome.

FAO and WWC (World Water Council) (2015), Towards a Water and Food Secure Future: Critical Perspectives for Policy-makers, White Paper, FAO and WWC, Rome, and Marseilles. www.worldwatercouncil.org/fileadmin/world_water_council/documents/publications/forum_documents/FAO_WWC_Water_and_Food_WhitePaper.pdf

Fischer, G. et al. (2007), "Climate change impacts on irrigation water requirements: Effects of mitigation, 1990–2080", *Technological Forecasting & Social Change*, Vol. 74, pp. 1083–1107.

Forzieri, G. et al. (2014), "Ensemble projections of future streamflow droughts in Europe", Hydrology and Earth System Sciences, Vol.18/1, pp. 85–108. DOI: http://dx.doi.org/10.5194/hess-18-85-2014

Gleeson, T. et al. (2012), "Water balance of global aquifers revealed by groundwater footprint", *Nature*, Vol. 488, pp. 197-200.

Hanasaki, N. et al. (2008), "An integrated model for the assessment of global water resources – Part 2: Applications and assessments", *Hydrology and Earth System Sciences*, Vol. 12, pp. 1027-1037.

Hanjra, M.E. and M.E. Qureshi (2010), "Global water crisis and future food security in an era of climate change", *Food Policy,* Vol. 35, pp. 365-377.

Hart, B.T. et al. (2003), "Ecological risk to aquatic systems from salinity increases", *Australian Journal of Botany*, Vol. 51/6, pp. 689-702.

Herring, S.C. et al. (eds.) (2015), "Explaining Extreme Events of 2014 from a Climate Perspective", Special Supplement to the *Bulletin of the American Meteorological Society,* Vol. 96, No. 12.

Hertel, T. W. and J. Liu (2016), "Implications of water scarcity for economic growth", OECD Environment Working Papers, No. 109, OECD Publishing, Paris. http://dx.doi.org/10.1787/5jlssl611r32-en

HLPE (2015), "Water for food security and nutrition", a report by the High Level Panel of Experts on Food Security and Nutrition of the Committee on World Food Security, Rome.

IFPRI and Veolia (2015), "The murky future of global water quality", White Paper, IFPRI and Veolia USA, Washington, DC.

Iglesias, A. and L. Garrote (2015), "Adaptation strategies for agricultural water management under climate change in Europe", *Agricultural Water Management*, Vol. 155, pp. 113–124.

Ignaciuk, A. et al. (2015), "Better drip than flood: reaping the benefits of efficient irrigation", *EuroChoices,* Vol. 14 /2, pp. 26-32.

Ignaciuk, A. and D. Mason-D'Croz (2014), "Modelling Adaptation to Climate Change in Agriculture", OECD Food, Agriculture and Fisheries Papers, No. 70, OECD Publishing, Paris. DOI: http://dx.doi.org/10.1787/5jxrclljnbxq-en

Jiménez Cisneros, B.E., et al. (2014), "Freshwater resources". in Field, C.B. et al. (eds.), *Climate Change 2014: Impacts, Adaptation, and Vulnerability: Part A: Global and Sectoral Aspects, Contribution of Working Group II to the Fifth Assessment Report of the Intergovernmental Panel on Climate Change*, Cambridge University Press, Cambridge, UK and New York, pp. 229-269.

Jongman, B. et al. (2014), "Increasing stress on disaster-risk finance due to large floods", *Nature Climate Change*, Vol. 4, pp.264–268. http://dx.doi.org/10.1038/nclimate2124

Lesk, C. et al. (2016), "Influence of extreme weather disasters on global crop production", *Nature*, Vol. 529, pp.84-87.

Le Kama, A. A., and A. Tomini (2013), "Water Conservation Versus Soil Salinity Control", *Environmental Modeling and Assessment*, Vol. 18/6, pp. 647-660.

MacLeod et al. (2015), "Cost-Effectiveness of Greenhouse Gas Mitigation Measures for Agriculture: A Literature Review", *OECD Food, Agriculture and Fisheries Papers*, No. 89, OECD Publishing, Paris, http://dx.doi.org/10.1787/5jrvvkq900vj-en.

Mateo-Sagasta, J. and J. Burke (2011), "SOLAW Background Thematic Study – TR08", FAO, Rome.

MEDDEE (Ministère de l'Écologie, du Développement Durable et de l'Énergie) (2012), "Hydrologie Souterraine – Synthèse », Report from the Explore 2070 project, Paris.

Mekonnen, M. M. and A. Y. Hoekstra (2016), "Four billion people facing severe water scarcity", *Science Advances*, Vol. 2/2, e1500323, DOI: 10.1126/sciadv.1500323

NASA Goddard Space Flight Center (2017), "Sea level: latest measurement April 2017- Satellite Data 1993-present", NASA, Pasadena, California. https://climate.nasa.gov/vital-signs/sea-level/ (consulted 1 September 2017).

OECD (2016), *Mitigating Droughts and Floods in Agriculture: Policy Lessons and Approaches*, OECD Studies on Water, OECD Publishing, Paris, http://dx.doi.org/10.1787/9789264246744-en.

OECD (2015), *Drying wells, rising stakes: Towards sustainable agricultural groundwater use*, OECD Studies on Water, OECD Publishing, Paris, http://dx.doi/10.1787/978926423870-en.

OECD (2014a), *Climate Change, Water and Agriculture: Towards Resilient Systems*, OECD Studies on Water, OECD Publishing, Paris, http://dx.doi.org/10.1787/9789264209138-en.

OECD (2014b), "Recommendation of the Council on the Governance of Critical Risks" [C/MIN(2014)8], OECD, Paris, France.

OECD (2013a), *Water and Climate Change Adaptation: Policies to Navigate Uncharted Waters*, OECD Studies on Water, OECD Publishing, Paris, http://dx.doi.org/10.1787/9789264200449-en.

OECD (2013b), *Water Security for Better Lives*, OECD Studies on Water, OECD Publishing, Paris, http://dx.doi.org/10.1787/9789264202405-en.

OECD (2012a), *OECD Environmental Outlook to 2050: The Consequences of Inaction*, OECD Publishing, Paris, http://dx.doi.org/10.1787/9789264122246-en.

OECD (2012b), *Water Quality and Agriculture: Meeting the Policy Challenge*, OECD Studies on Water, OECD Publishing, Paris, http://dx.doi.org/10.1787/9789264168060-en.

OECD (2011), *Managing Risk in Agriculture: Policy Assessment and Design*, OECD Publishing, Paris, http://dx.doi.org/10.1787/9789264116146-en.

OECD (2010), *Future Global Shocks*, OECD Reviews of Risk Management Policies, OECD Publishing, Paris, http://dx.doi.org/10.1787/9789264114586-en.

OECD (2009), *Managing Risk in Agriculture: A Holistic Approach*, OECD Publishing, Paris, http://dx.doi.org/10.1787/9789264075313-en.

OECD (2007), *Effective Targeting of Agricultural Policies: Best Practices for Policy Design and Implementation*, OECD Publishing, Paris, http://dx.doi.org/10.1787/9789264038288-en.

OECD/ C. Ringler (2011), "Effect of Reduced Water Supplies on Food Production Economies", in *Challenges for Agricultural Research*, OECD Publishing, Paris, http://dx.doi.org/10.1787/9789264090101-6-en.

Orr, S., et al. (2012), "Dams on the Mekong River: Lost fish protein and the implications for land and water resources", *Global Environmental Change*, Vol. 22, pp. 925–932.

Pachauri et al. (2014), "Climate Change 2014-Synthesis Policy Report: Summary for Policymakers", Intergovernmental Panel on Climate Change (IPCC), Geneva.

Pauw, K. et al. (2010), "The economic costs of extreme weather events : a hydro-meteorological CGE analysis for Malawi", *Environment and Development Economics*, Vol 16, pp.177-98.

Pfister, S. and P. Bayer (2013), "Monthly water stress: spatially and temporally explicit consumptive water footprint of global crop production", *Journal of Cleaner Production*, Vol. 30, pp. 1-11.

Pimentel, D. (2005) "Environmental and Economic Costs of the Application of Pesticides Primarily in the United States", *Environment, Development and Sustainability*, Vol. 7/2, pp. 229-252.

Ringler, C. et al. (eds.) (2010), *Global Change: Impacts on water and food security, Water Resources Development and Management*, Springer, New York, United States.

Sadoff, C. et al. (2015), "Securing Water, Sustaining Growth", Report of the GWP/OECD Task Force on Water Security and Sustainable Growth, Oxford University, Oxford. https://www.water.ox.ac.uk/wp-content/uploads/2015/04/SCHOOL-OF-GEOGRAPHY-SECURING-WATER-SUSTAINING-GROWTH-DOWNLOADABLE.pdf

Schaible, G.D., and M. Aillery (2016),"Challenges for US irrigated agriculture in the face of emerging demands and climate change", in Ziolkowska, J.R. and J. M. Peterson (eds.), *Competition for Water Resources: Experiences and Management Approaches in the U.S. and Europe*, Elsevier Publishing, Amsterdam,

Schlosser et al. (2014), "The future of global water stress: An integrated assessment", Report N. 254, MIT Joint Program on the Science and Policy of Global Change, MIT, Cambridge, MA. http://globalchange.mit.edu/files/document/MITJPSPGC_Rpt254.pdf

Schoengold, K. and D. Zilberman (2007) "The economics of water, irrigation, and development", in and R. Evenson and P. Pingali (eds.), *Handbook of Agricultural Economics*, Vol. 3, pp. 2933-2977. doi:10.1016/S1574-0072(06)03058-1

UNEP (United Nations Environment Program) (2016), "Food Systems and Natural Resources", a report of the Working Group on Food Systems of the International Resource Panel, written by H., Westhoek, et al., UNEP, Nairobi. www.unep.org/resourcepanel/knowledgeresources/assessmentareasreports/food

Veolia Water and IFPRI (International Food Policy Research Institute) (2011), "Sustaining growth via water productivity: 2030/2050 scenarios", Final Results Document, Veolia Water, USA.

Vörösmarty, C. J. et al. (2000), "Global Water Resources: Vulnerability from climate change and population growth", *Science,* Vol. 289, p. 284.

Wada, Y. et al. (2011), "Modelling global water stress of the recent past: on the relative importance of trends in water demand and climate variability", *Hydrology and Earth System Sciences*, Vol. 15, pp. 3785-3808.

Wada, Y., et al. (2010), "Global depletion of groundwater resources", *Geophysical Research Letters*, Vol. 37, p. L20402. DOI:10.1029/2010GL044571.

WEF (World Economic Forum) (2015), "Global Risks 2015", 10th report, WEF, Geneva.

WWAP (United Nations World Water Assessment Programme) (2015), "The United Nations World Water Development Report 2015: Water for a Sustainable World" UNESCO, Paris.

WWAP (2012) "The United Nations World Water Development Report 2012: Managing water under uncertainty and risk", UNESCO, Paris.

2030 WRG (2030 Water Resources Group) (2009), "Charting our Water Future", International Financial Centre, Word Bank Group, Washington DC.

Chapter 2

Defining and identifying water risks for agriculture

This chapter defines the key characteristics of a hotspot approach to agriculture water risks. It then applies this approach at a global level, using data from the literature, to identify future water risk hotspot countries for agriculture production. The evidence points to the People's Republic of China, India and the United States as the leading agricultural producing countries most likely to be impacted. Specific water risks within these countries, in the identified key agriculture production regions of Northeast China, Northwest India and Southwest United States, are reviewed.

The statistical data for Israel are supplied by and under the responsibility of the relevant Israeli authorities. The use of such data by the OECD is without prejudice to the status of the Golan Heights, East Jerusalem and Israeli settlements in the West Bank under the terms of international law.

Key messages

The "hotspot" approach focuses attention and action on locations where the risk is highest relative to other locations due to higher hazards, exposure or vulnerability within a broader approach to risk management. This approach presents clear advantages for policy makers. It lowers the cost for an achievable result, helps prevent the diffusion of risks, and focuses attention and efforts on specific regions. It can be applied at virtually any scale for individual or multiple risks, and can account for all dimensions of water risks in a limited number of spots.

This approach is particularly well suited to agriculture water risks given the local specificities and dynamics associated with agriculture and water that make it difficult to propose generic approaches to risks. The hotspot approach can also help control pollution and support customised adaptation to climate change.

Application of the hotspot approach is only effective under certain circumstances. There are three main conditions that must be met: (1) risks should be non-uniform at the national level; (2) high-risk regions should be well-defined and bounded; and (3) hotspot assessments should be supported by robust information and data.

Employing the hotspot approach to agriculture water risks involves two steps: defining agriculture water risks and determining a threshold to identify what constitutes a hotspot. Both steps will vary depending on the scope, scale and time horizon, as well as the level of information.

The hotspot approach was applied to future agriculture production at the global level – using a combination of a literature review on water risks and existing agriculture projection – to determine countries at high risk. This exercise identified China, India, and the United States as the top three water risk hotspot countries for agriculture production. While many other countries and regions are expected to face high water risks that will affect their agriculture production, these three countries are distinguished by the fact that they concentrate high levels of water risks and projected high shares of global agriculture production.

Three key agriculture production regions within these countries are expected to face particularly high water risks: Northeast China, Northwest India, and the Southwest United States. These regions face similar water issues such as low and variable surface water supplies, groundwater depletion, and expected increased demand from other sectors. Water quality issues are also prevalent in Northeast China and could arise in Northwest India.

2.1. Rationale and conditions for a robust hotspot approach

The "hotspot" approach focuses attention and action on locations where the risk is highest relative to others due to higher hazards, exposure or vulnerability, within a broader approach to risk management.[1] The management of risks can operate at different levels. Public policy can be set at the international, national or sub-national levels when considering broad types of risk, and prioritisation may help advance a strategy or investment among multiple risk dimensions. In practice, all broad applications, prioritisations, or hotspot approaches may act as a complement for different types of risks in response to general or critical risks. This approach does not aim to replace overall management of water resources and the necessary broader approach to risks, but aims to act as a complementary policy for higher effectiveness when faced with critical risks.[2]

The hotspot approach has multiple advantages when risks are geographically concentrated. First, it allows for gains in efficiency by lowering the cost of an achievable result. Financing measures that prioritise actions to address the most critical risk areas deters investment in lower priority risks that have limited beneficial outcomes. Second, it may prevent the diffusion of risks; advanced targeting is often a key component of pollution or sanitary damage control to prevent wider impact. Third, when applied *ex ante* to vulnerable areas, it can help focus attention and efforts on uncertain futures (FAO, 2015a). Fourth, since it is an approach based on relative levels of risks, it can be applied at virtually any scale for multiple risks. Lastly, the focus on hotspots enables an all-encompassing response to risks, taking account of all dimensions of water risks in a limited number of hotspots, from water scarcity to water quality and variability. This approach may serve as a learning process to gradually address a greater number of areas with lower risks.

In the context of this study, the hotspot approach identifies specific locations that are most likely to be subject to agriculture water risks in the future, because they are, or will be, significant agriculture producing regions and are they expected to be under water constraints. For the purposes of this report, water risks are defined as the combination of different types of water constraints: insufficient water, water abundance, water quality impairment, and water-related catastrophic events (see Annex 2.A1). The objective is to identify and put in place effective responses that are specifically adapted to locally important risks for agriculture. While some of the responses may include options that are not locally specific, such as varietal breeding or efficient irrigation systems, the intensity and combination of responses need to match the scale of the local challenges (Chapter 4).[3] For instance, agriculture regions with groundwater intensive use that leads to multiple environmental externalities, as observed in California, require a combination of advanced responses, from increased information to collective action, and the use of regulatory and economic instruments, that are not needed in areas not facing the same problems (OECD, 2015a).

In this context, the hotspot approach may have several additional benefits:

- A narrow scope can help target water risk mitigation policies in areas where impacts will most likely be critical. As agriculture activities vary geographically, the sector's vulnerability to and impact on water risks will also differ locally. This is particularly the case for large countries or regions with a diversity of climatic conditions and agriculture systems.
- This approach can mitigate water quality risks, whether associated with point source or diffuse pollution, spread via surface waterways and groundwater bodies. Targeting is a key recommendation for effective water quality management (OECD, 2013), including in agriculture (OECD, 2012).[4]
- The hotspot approach can also be a means to design high return, locally-customised climate adaptation plans that account for productivity objectives. The efficiency and effectiveness of public intervention can be enhanced by knowing more about where and how to act (Ignaciuk, 2015).
- It allows for policy actions that consider the cumulative effects of multiple hotspots risks materialising simultaneously. Projections can present climate-related water impacts affecting demand and supply in different regions simultaneously. For example, the Russian heat wave and the Pakistan flood that occurred simultaneously in 2010 were climate-related (Lau and Kim, 2012), and likely had consequences on global food markets.

- The proposed approach can be employed at different agriculture, geographical or administrative levels as needed. Targeting for policy purposes will generally be more relevant at the national level, but state or province level may also be more appropriate in countries with federal government systems. Watersheds can be set as unit subject to prioritisation, or targeting can also be done within a specific watershed.

It should be noted that applying future water risk hotspots for agriculture does not necessarily imply that current allocation of efforts is not effectively suited to respond to water risks. Instead it encourages further policy efforts in this direction, especially considering future water risks that may not be currently addressed, either because they do not cover the same area, or because the risks are likely to further increase in areas facing current risks. Baseline projections of water risks show that the cost of no additional action can be significant (e.g. Chapter 1), the hotspot approach essentially aims to focus on areas where this cost is expected to be the greatest.

Conditions for the beneficial application of the hotspot approach

The relevance of the hotspot approach relies on the presence of a non-uniform distribution of risks in space, time, or hazard intensity (see Annex 2.A1 for definitions). For example, a large aquifer could be at risk of complete depletion, but the timeframe for depletion is unknown, if it occurs, the impact could vary depending on type of user and location, including users across country borders. Overall, the water cycle is subject to major non-uniformities, from variations in climate and precipitation to interactions with continental landscapes, ocean currents, local temperature, and human activities; "Availability of water is very different across geographical regions, both in terms of rainwater, and of surface and ground water. Therefore, water availability needs to be considered at regional, national and local levels" (HLPE, 2015).

The usefulness of this approach is also determined by the presence of well-defined risk peaks, or regions of relatively high risks, that are sufficiently narrow but still significant in scope (Figure 2.1). A wide distribution of low risks reduces the effectiveness of the hotspot approach. For example, if salinity affects a whole continent with a relatively low overall impact, a hotspot approach may not be as useful as compared to a limited portion of a highly productive agriculture coastal area at risk from salinity. At the same time, punctual discrepancy in risk profiles (outliers) should not qualify as peaks, and a sufficiently large area or activities needs to be affected for the corresponding area to be considered a hotspot. These conditions will depend on the water problem and results of the risk assessment process.

Effectively targeting risk requires sufficient knowledge and information. Hotspot assessments must be sufficiently robust to avoid potentially costly mistakes, particularly when considering future risk hotspots. Missing a hotspot—a statistical type II error—could be costly. A high cost could also be associated with investing in institutional efforts, funding or regulatory actions on a site or region that turns out not to be at risk (a statistical type I error). Avoiding such mistakes requires gathering sufficient water risk data and providing trust-worthy estimations of risks using multiple hypotheses, scenarios and sources for validation when and if possible.

Finally, the hotspot assessment must be undertaken at the appropriate scale. Too large a scope may neglect important local problems. Similarly too narrow a scope may result in excessive attention to specific problems to the detriment of important issues observable at a broader scale. There may be trade-offs to consider in national versus regional hotspot assessment.

Setting the proper threshold to distinguish hotspot areas from non-hotspot areas requires accounting for these conditions. There is an inherent trade-off between the selected threshold level, which increases efficiency and impact, and the coverage of the assessment (as illustrated in Figure 2.1). Under conditions of perfect knowledge, such a trade-off could be resolved via a cost-benefit analysis, but under conditions of uncertainty on hotspots, decision makers must decide the extent to which savings must be sacrificed to avoid possible mistakes.

Figure 2.1. Selecting a threshold for risk hotspots requires balancing focus and coverage

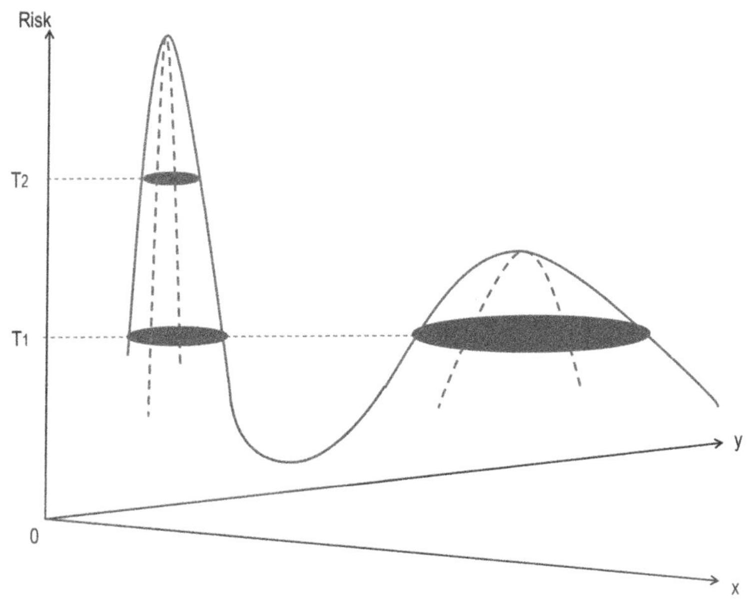

Note: The curve shows the evolution of risks, defined as exposure multiplied by expected hazard, in the (x,y) space. Cutting at level T_1 allows including two hotspot areas of potential importance, but could create significant costs and lower efficiency, cutting at T_2 increases focus, reducing actions to one hotspot with more likely effective response but potentially insufficient scope.

Source: Author's own work.

2.2. Assessing water risk hotspots for agriculture production: Methodology and application at the global level

Main steps to identify water risk hotspots for agriculture

There are two steps to identifying agriculture hotspots (Annex 2.A1).

- *Definition.* Defining future water risk hotspots for agriculture requires the measurement of water risks affecting agriculture and plausible ways to project these risks in the future. There are multiple indices used to measure water risks, each with its own advantages and possible limitations. Choosing the right index will depend on the type of risks and availability of data and/or modelling means (Annex 2.A1). Identifying hotspots also requires credible agriculture projections. Both water risk and agriculture projections will depend on the specific time horizon and geographic scope. The two types of projections need to be combined or integrated: agriculture conditions (in the absence of risks and responses) and water risk assessments.

- *Setting appropriate thresholds to define hotspots.* There are several options to consider depending on the degree of detail taken in the first step of the assessment,. Where the future water risks and agriculture projections are well known, thresholds can be determined on the basis of the distribution of risks on agriculture (or the estimation thereof). Under partial or incomplete information on future water risks hotspots for agriculture, the objective is to look for regions with a consistently higher level of projected agriculture water risks (combined water risks and agriculture importance) relative to other regions based on available evidence. Where critical information on water risks or agriculture is unavailable, or where information focuses on a limited area, a hotspot approach may not be recommended.

Annex 2.A1 provides a more complete explanation on the methodology used to define water risks, identify agriculture water risk hotspots, including regional and national examples of applications of water risk hotspot approaches from the European Union and the United States, Australia, Switzerland and New Zealand. The next section explains the method used to identify the globally-significant agriculture-producing countries subject to future high water risks.

Application: Searching for globally significant water risk hotspots for agriculture production

The objective of the present application is to identify countries where agriculture production is projected to face the highest water risks and have significant global impact. In this application, agriculture production levels certainly matter but only as far as such production will face high water risks. This implies that countries with high water risks and low projected production or countries with high agriculture production and water risks may not be identified as hotspot compared to others. In contrast, countries that concentrate production and water risks criteria have a global significance in that there response to water risk may affect global markets and therefore a wider range of countries. The water risks considered include shortage, excess, and water-quality related, as defined in Annex 2.A1.

Applying a hotspot approach to water risks at the global level is a difficult exercise. It requires extensive inquiries as well as a process to address multiple uncertainties. The models used need to pull from sufficiently well-calibrated data in all regions of importance. There may be significant uncertainties associated with assumptions for both water and agriculture. On the water risk side, climate change effects and demand expectations from agriculture and other sectors may be uncertain (Buckle and Mactavish, 2014; OECD, 2014a, see Box 2.1). For the agriculture sector, the critical issue is to assess the future of the sectors in the absence of risks —i.e. establishing a credible counterfactual— to ensure that risks within the hotspots are indeed important.

Box 2.1. Projecting water risks associated with climate change: A confluence of uncertainties

There is clear value in trying to project climatic conditions and to use them in hydrologic assessments. They provide insights on broad trends that can be useful if not critical to water managers operating in areas under growing water stress. But these projections also face a number of challenges and uncertainties.

Multiple researchers note that, contrary to temperature, simulating changes in water cycles is challenging. Like any climate projection, they first face multiple uncertainties stemming from: (1) scenarios of future greenhouse gas emissions by integrated assessment models; (2) translation of greenhouse gas emissions scenarios into atmospheric concentrations and forcings; (3) evaluation of the effects of these forcings on climate by global climate models (GCMs); (4) downscaling and bias-correcting the output of the GCMs; and (5) translation of climate change projections into impact projections by impact models, e.g. hydrological or vegetation models (Döll et al. 2015). National rainfall projection uncertainties, in particular, require periodical bias corrections between simulated and historical data to be of use. Inherent uncertainty in computing freshwater-related hazards adds to this picture, as there is generally scarce information on the state of freshwater systems (Ibid.).

These uncertainties render the results of quantified simulations potentially unrealistic. Döll et al. (2015) argue that these exercises provide probabilities of possible future water risks and hazards. There is some evidence to suggest that even with broadly based results, the recommendations they provide may have little value. A review of the application of 28 World Bank studies using climate projection models found that they are often used as a backdrop for urging the adoption of "no-regret" actions, and rarely for quantitative decision making on options". The review concludes that climate model information has generally been unable to inform quantitative decision making in the surveyed sample [...] over half of the studies recommended low-regret adaptation options that do not depend on climate projections, and roughly one-quarter did not recommend adaptation options" (IEG, 2011).

The use of models for broad hotspot mapping is also questioned by researchers. De Sherbinin (2014) argues that in many cases data-driven maps show patterns that would have been identified in an expert assessment approach or based on a broad understanding of past patterns. In particular, regions that have the lowest levels of economic development are typically found to be most at risk in global hotspots mapping assessments" (Ibid.).

Source: Döll et al. (2015); de Sherbinin (2014); IEG (2011).

There are also important caveats to this particular application of the hotspot approach, which focuses on water risk for agriculture production at the global level.

- It relies entirely on secondary data and available literature and therefore represents an example of hotspot determination based on an incomplete assessment. The hotspot application is based on a finite sample of studies on global water risk that does not aim to be fully comprehensive. The dataset derived from the literature uses multiple types of models that are not always comparable.[5] While combining results from multiple approaches helps provide some robustness that single simulation studies may lack, the results of the proposed assessment must be viewed as a second-best assessment of future water risks in agriculture.
- It provides a snapshot of evidence that can evolve over time. New areas and countries may become hotspots tomorrow with unpredictable changes in climate or water use.
- The present exercise is done at a global level with countries as the primary unit of analysis because of data and analysis limitations (superposing precise geographic data would require compiling all data from the large range of studies, which are not available). This means that the size of countries matters in the exercise given the relationship and impact this factor has on agriculture production levels. At the same time water risk level do not follow country borders and may be highly concentrated in areas with small or larger countries.[6]
- The emphasis of the assessment is to look at future water risks; therefore, most studies assess water risks relative to the current risk status of countries. As a consequence countries with high water risks today but projected favourable future climatic conditions may not be singled out as subject to high risk in the assessment.
- The proposed application, which focuses on agriculture production, is more likely to consider regions with significant importance for global food security than regions that have greater local and regional food security issues. Multiple regions and countries where local agriculture is critical for the local populations, face high water risks, and yet may not be identified as hotspots. The methodology put forth in this assessment accounts for the fact that agriculture products are increasingly traded and international trade could therefore act as a tempering mechanism for agriculture water risks of local importance. In contrast, in the case of high risks for globally important agriculture countries, water constraints could lead to global supply imbalances and much larger market effects with possible food security implications. Integrated modelling that accounts for water, climate, and food security risks could help even if the hypotheses and scenarios may limit the robustness of the results.
- Vulnerability is not as a core criterion for hotspot selection given the lack of consistent data on future vulnerability to water risks for agriculture and the emphasis on risks that will have an impact on global food production, Instead, the methodology prioritises the likelihood and expected agriculture production impacts.

The assessment relies on the geographical decomposition of results from 64 global-level studies with water risks measurements (see Annex 2.A2 for details). These studies assess the different types of water risks associated with climate change and/or demand projections, focus on surface water and/or groundwater in the current, medium term or long term, all at the global scale. Most studies focus on water risks for all sectors and not specifically agriculture, and a few studies look at vulnerability. Most focus on likelihood and intensity of impacts (measured in different ways, see Annex 2.A1). To the extent possible, the assessments use business as usual or no action scenarios as opposed to scenarios with simulated responses (e.g. water risk adaptation or mitigation of risk).

The hotspot approach uses the frequency in observations considering that a region is of high risk as a primary metric for eligibility. Countries are considered—and accounted for in the computation of this frequency— if they are categorised in the high or highest risk categories or if they are identified in any report as facing the most severe water risks. Availability of data across countries is a limiting factor of this indicator, which could lead to inconsistency in results with respect to level of risk by country. However, given the global focus of the hotspot approach, the overwhelming majority of reviewed studies uses genuine global simulation models that span through all continents, preventing a systemic bias. Furthermore, the consistency across studies that frequency measures may reflect research preference for types of modelling (model, scenarios etc.), but does not guarantee that such methods are the best. At the same time, it reduces the possible bias associated with using one single model.

On the agriculture production side, the methodology considers a set of widely used commodities in the medium run and the longer run, presuming absence of water or climate risks (counterfactual scenario). These countries are or will become agriculture hotspots because their production sustains a significant share of the world's aggregate supply. This is done by collecting estimates of production and exports for coarse grains, rice, wheat, oilseeds, sugar, cotton, dairy, and beef from the OECD-FAO Agricultural Outlook 2015 (OECD/FAO, 2015) for 2024, and the same commodity from the baseline scenario of IFPRI's IMPACT model for 2050 (Robinson et al., 2015). The fruit sector was also looked at for 2050 as it may be affected by water (but is not specifically separated in the AgLink-Cosimo model). The indicators for agriculture weights are defined as the average shares of production for the selected commodities and time horizons.

Future water risk hotspots are then selected by cross-referencing information on water risk with information on agriculture production (presence of a water risk as indicated with the frequency of high risk index and high share of production). No explicit threshold is assumed a priori, as the determination is completed by relative comparison. Those countries or regions with consistently higher risks and higher agriculture market shares in at least some of the key commodities are potential candidates for future water risk hotspots for agriculture.

2.3. China, India, and the United States concentrate global agriculture production water risks

The determination of water risks was based on 118 observations of water risks (current or future), from 100 individual analyses, coming from 64 publications of global water risks (including risks in agriculture), which are listed in Annex 2.A2. Table 2.1 shows the distribution of observations; the overwhelming majority focuses on water scarcity risks.

Table 2.1. Distributions of observations from the literature review

	Risk of shortages	Risk of excessive water[1]	Risk of climate variability[2]	Water quality risks	Total
Future	59	20	4	3	86
Current	28	3	1	0	32
Total	87	23	5	3	118

1. Includes risks of flooding generated by sea level rises.

2. This category regroups observations that capture the probability of extreme events lump together. Observations for extreme floods or droughts were counted under shortage and excess, respectively.

Source: Author's assessment based on the reviewed literature.

The location of hotspots was most often determined directly from maps of risks, using tables provided by the publications studied, and/or by looking at the authors' assessments and analysis of the most severe risks. These results were reported by indicator variables – a value of 1 assigned if the risk is prevalent, and a value of 0 assigned for non-prevalent risk – for 142 countries.[7] Results reported in publications at the regional level (e.g. Middle East or North Africa) were then accounted for at the country level for the corresponding countries in the region (e.g. Algeria was allocated a 1 if North Africa was considered a hotspot in a study).

The most severe water risks were found in two bands of countries largely in the north and south subtropical zones (Figure 2.2).[8] For countries that are subject to high risks in at least 30% of observations, results were decomposed into four categories of risks: shortages, excess, variation (probability of extreme events), and quality (Figure 2.3). Thirty countries are in this set. The leading three countries – China, the United States and India – were found to be at high risk in over 55% of the measurements.[9] Fifteen countries follow, with the proportion of high risks observations between 40% and 50%, mostly from the Mediterranean region. The remaining 16 countries are relatively spread out and diverse. Ten OECD countries feature in this list: Australia, Chile, France, Greece, Israel, Italy, Mexico, Spain, Turkey, and the United States.

Figure 2.2. China, India and the United States are expected to face the most water risks

Frequency of observations, listing countries as subject to high or very high future water risks

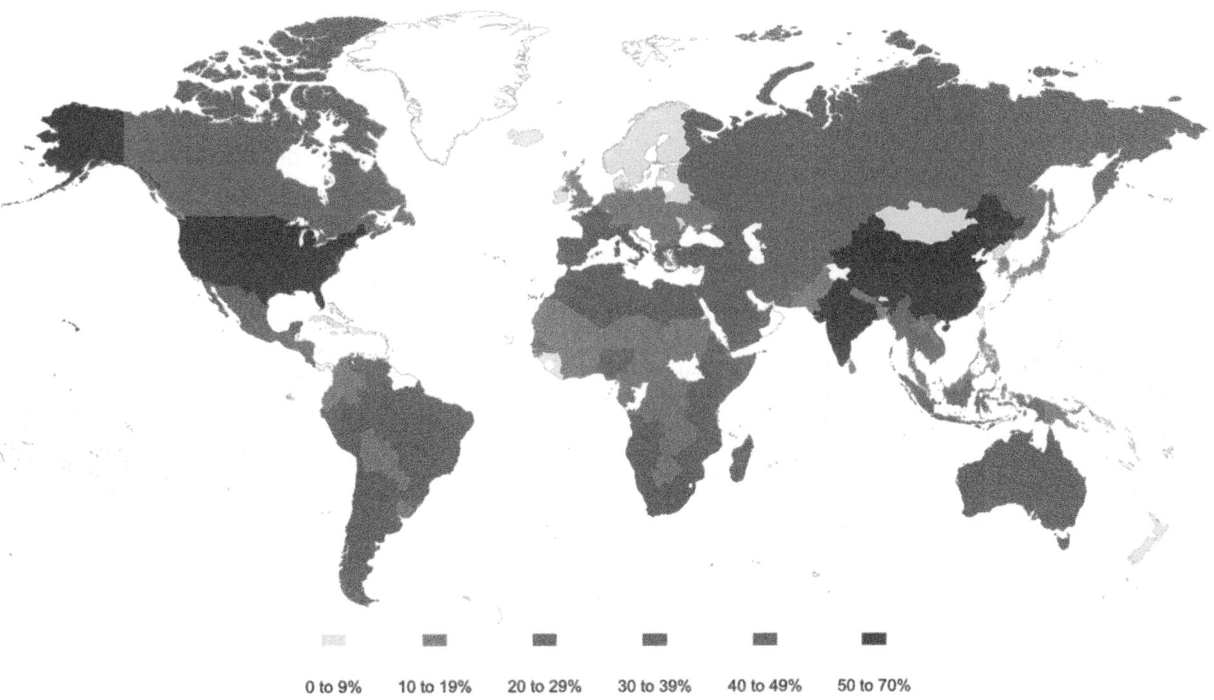

Source: Review of 64 publications, accounting for 142 countries.

Figure 2.3. Proportion of severe water risk observations for the leading countries in the reviewed literature

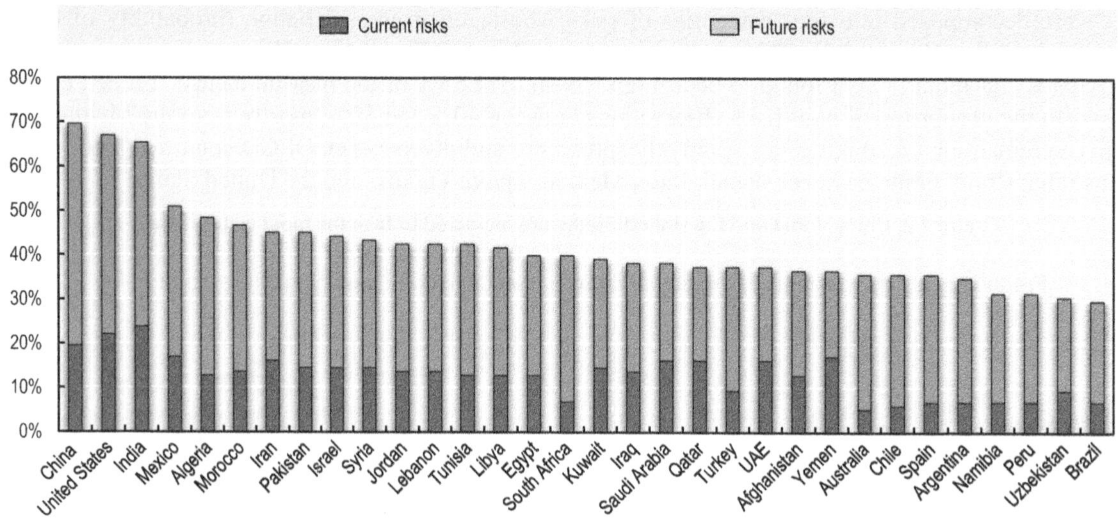

Source: Derived from an analysis of 64 studies.

Comparing observations of current and future risks (Figure 2.4) helps distinguish four groups of countries: the three leading countries (China, India and the United States) featuring the highest current and future risks; a second group of countries (located mostly in North Africa and the Middle East) with high current and future risks; a third group of countries with moderate to high risks (that includes Mediterranean, Latin American and Southern African countries); and a large group of countries subject to comparatively lower future water risks (< 30% in both dimensions).

Most countries that have high indicators for current water risks have relatively lower indicators for future water risks (they are located under the median), and conversely most countries with lower indicators for current water risks have higher indicators for future water risks (above the median).[10]

Looking at the type of risks (Figure 2.5), if most countries are subject to risks of shortages, the leading three countries (China, the United States and India) are subject to the three main risks (shortage, excess and quality). More broadly, Table 2.2 shows that China, India and the United States feature among the top listed countries in multiple categories of risks.

Figure 2.4. Countries with lower current water risks may face relatively higher risk

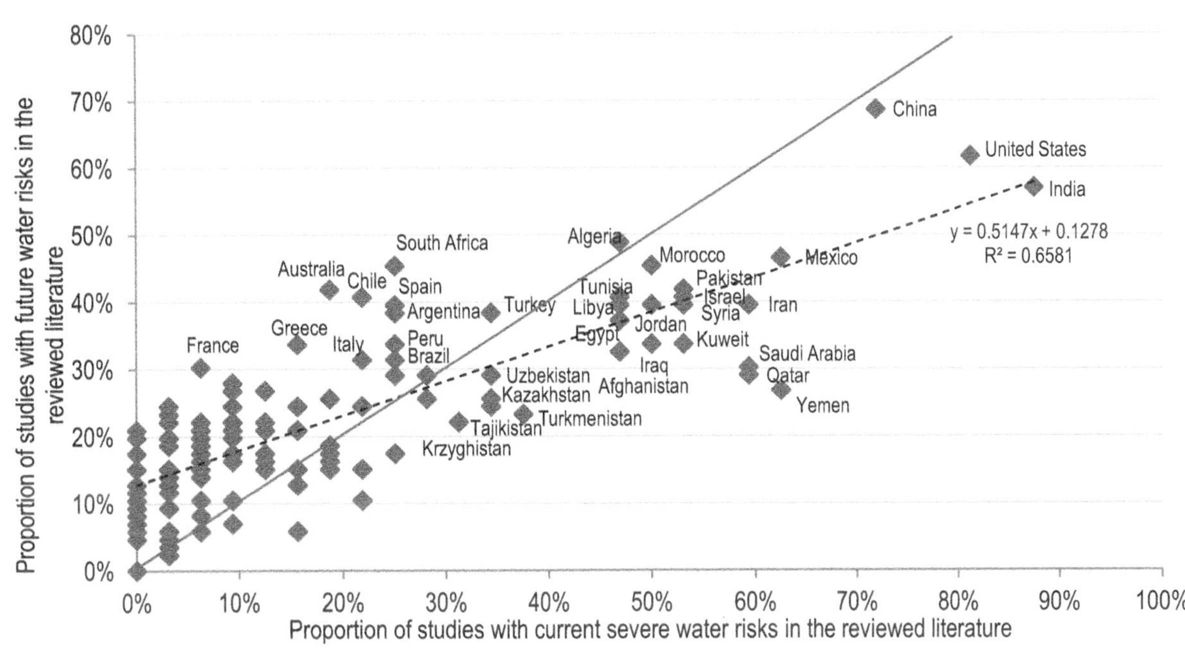

Note: Shares of severe water risk future and past observations, across reviewed studies.
The continuous line represents the median (y=x), the dashed line is a linear regression.

Source: Derived from the analysis of 64 studies.

Figure 2.5. Proportion of severe future water risks, by category, reported in the reviewed literature

Only countries with overall proportion above 30% are listed

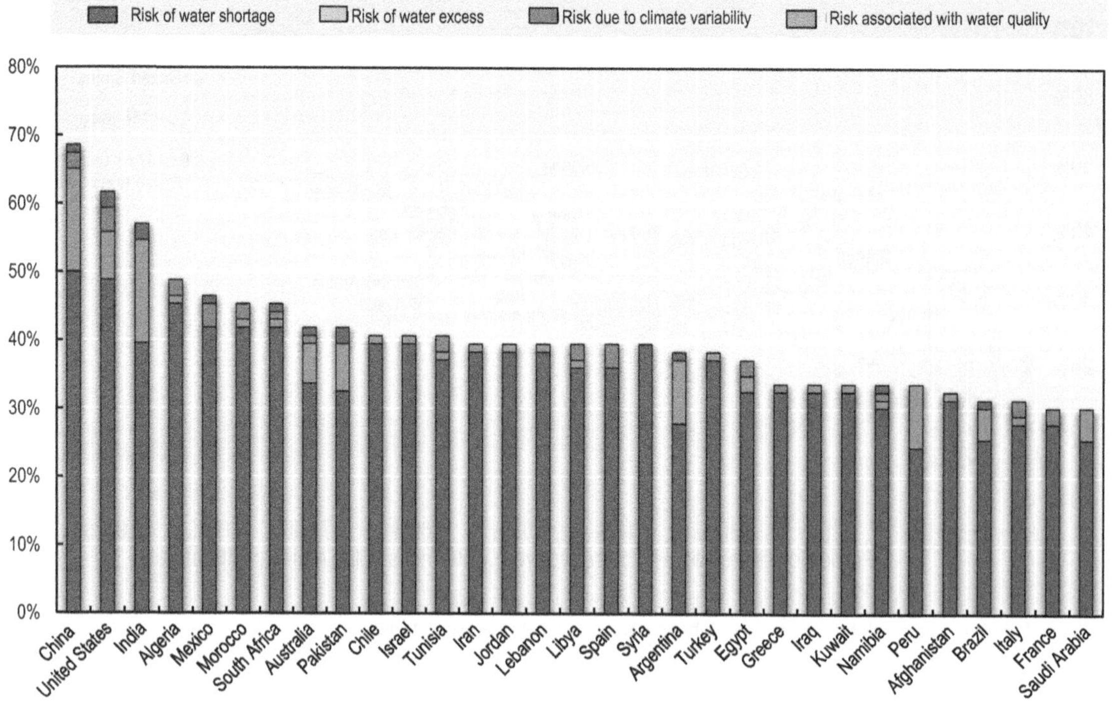

Source: Derived from the analysis of 64 studies.

Table 2.2. China, India and the United States lead the rankings for different types of water risks

Future water risks aggregate index	Future risks of water shortages	Future risks of excess water	Future risks of variability	Future risks of water quality
1. China (69%) 2. United States (62%) 3. India (57%)	1. China (73%) 2. United States (71%) 3. Algeria (66%)	1. China, India (65%) 3. Cambodia, Indonesia, Myanmar, Viet Nam (55%)	1. 12 countries including the United States (75%)	1. India, Pakistan, United States (67%)

Source: Review of 64 studies.

Looking ahead at the agriculture production for 2024 and 2050, Brazil, China, India and the United States account for about 50% of the average global production (Figure 2.6). These large countries consistently lead production rankings across almost all categories of products, whether currently or in 2024 and 2050 baseline projections. The four countries that follow — Argentina, Indonesia, Pakistan, and the Russian Federation — are projected to have significant production shares in at least a few markets for one or the other model. The twelve remaining countries appear either to have strong specialisations (e.g. Thailand for rice) or a non-negligible contribution in several markets (Ukraine).

Results from the two model projections (AgLink-Cosimo and IMPACT) are reconciled by taking average shares to represent medium-term projections (2024 from AgLink-Cosimo and 2050 from IMPACT) across commodities, keeping the average shares from IMPACT for 2050 with or without fruits to capture the longer term. Differences across models can be explained by the models' differences in structure, projections and hypotheses. Furthermore, some countries are only covered in one of the databases.

Figure 2.6. Average production shares of major agriculture commodities for the 20 largest producers, 2024 and 2050

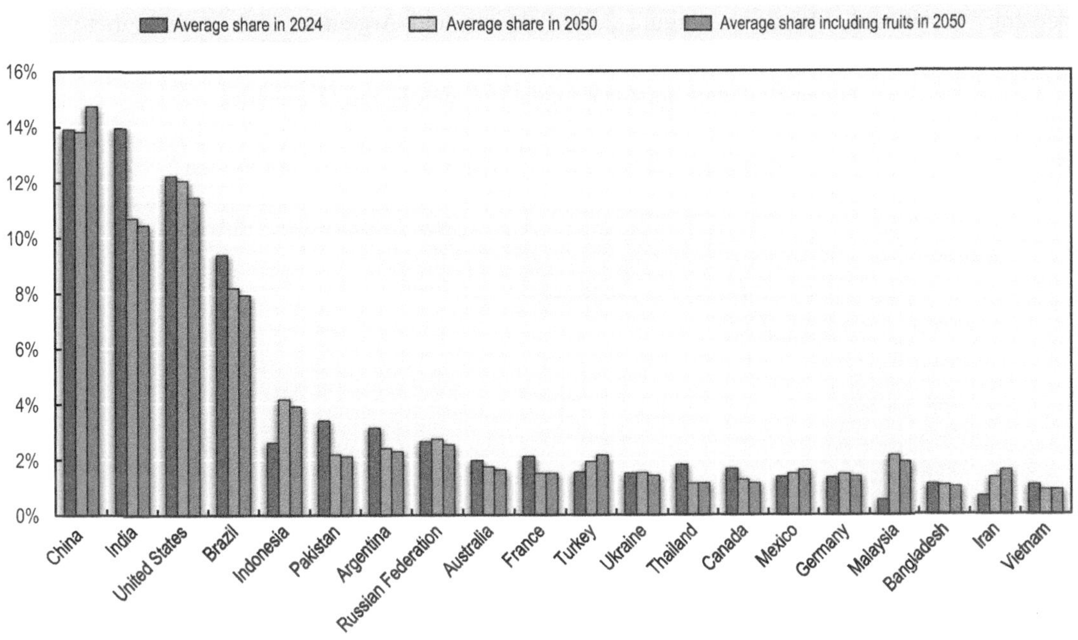

Note: Eight commodities are included: coarse grains, rice, wheat, sugar, cotton, oilseeds, dairy, and beef.
Source: Derived from baseline projections from the AgLink-Cosimo model for 2024, and from the IMPACT model for 2050.

The index representing future water risks for agriculture is computed by multiplying the proposed agriculture and water risk indicators, i.e. the proportion of measures reporting water risk in the future by the average shares of agriculture production (average of 2024 and 2050 as core, 2050 with the same commodities, and 2050 with fruits as addition), and a factor of 100. As such, the index can be interpreted as the expected share of overall global production of the selected commodities likely to face high water risks in each country in the medium to long run under no adaptation action. This index is computed for 77 countries that are the most significant agriculture producers of each commodity, representing altogether over 86% of total projected 2050 production of each commodity. It also includes countries of the OECD, ASEAN, and the Mediterranean region. The results are shown in Figure 2.7 for the top 15 countries, in Figure 2.8 comparing the leading countries to the aggregate index for selected regions, and in Figure 2.9 globally (OECD results are shown in Annex 2.A2, Figure 2.A2.1).

Figure 2.7 shows that three countries stand out from the analysis: the United States, China and India. These countries are expected to remain the leading international agriculture producers but also rank highest in terms of projected water risks. Their land and population scope may contribute to this ranking, although this is not sufficient to explain the high agreement across studies on the presence of future water risks (55% to 70% of observations report severe water risks of different kinds). To ensure that the size effects does not dominate the diagnostic, regional rankings for three multi-country regions with large shares of agriculture production and significant water risks were computed for comparison; the 14 Mediterranean coastal countries (which includes countries from North Africa, Southern Europe and the Near-East), ASEAN (Southeast Asia) and 21 European Union countries members of the OECD.[11] As shown in Figure 2.8, none of these regional groups exceeds half of the index of each of the three top countries. On this basis, and acknowledging the previously listed limitations of this global country-level analysis, China India and the United States are identified as future water risk hotspots for agriculture production.

Brazil and the Mediterranean region as a whole may be considered secondary future water risk hotspots for agriculture production (Figures 2.7 and 2.8). Despite its low rank in water risk in Figure 2.5, Brazil is projected to remain a leading producer of multiple commodities. All 14 Mediterranean countries are identified among the countries most exposed to high water risks (Figure 2.5), but even combined together, they represent a much lower share of agriculture production than Brazil, thereby reaching similar scores on Figure 2.8. Focusing on national differences (Figure 2.7), Pakistan and Argentina also score high for agriculture water risks, as they combine large national agriculture productions and significant water risks.

Figure 2.7. Future agriculture water risk indices, top 15 countries

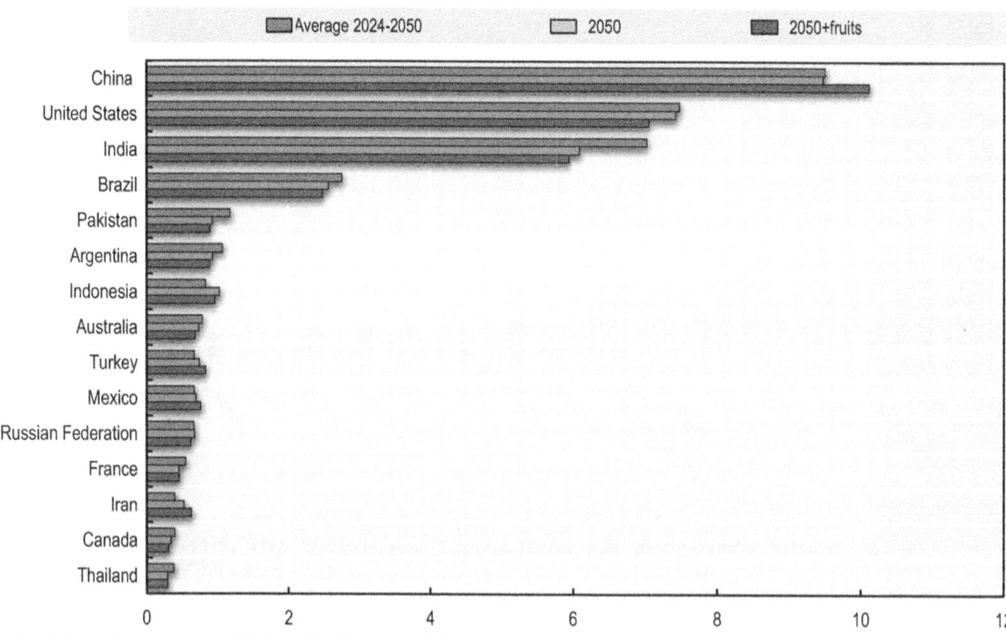

Source: Derived from the analysis of 64 publications, and AgLink-Cosimo and IMPACT projections.

Figure 2.8. Future agriculture water risk indices for the leading four countries and for four regions facing high water risks

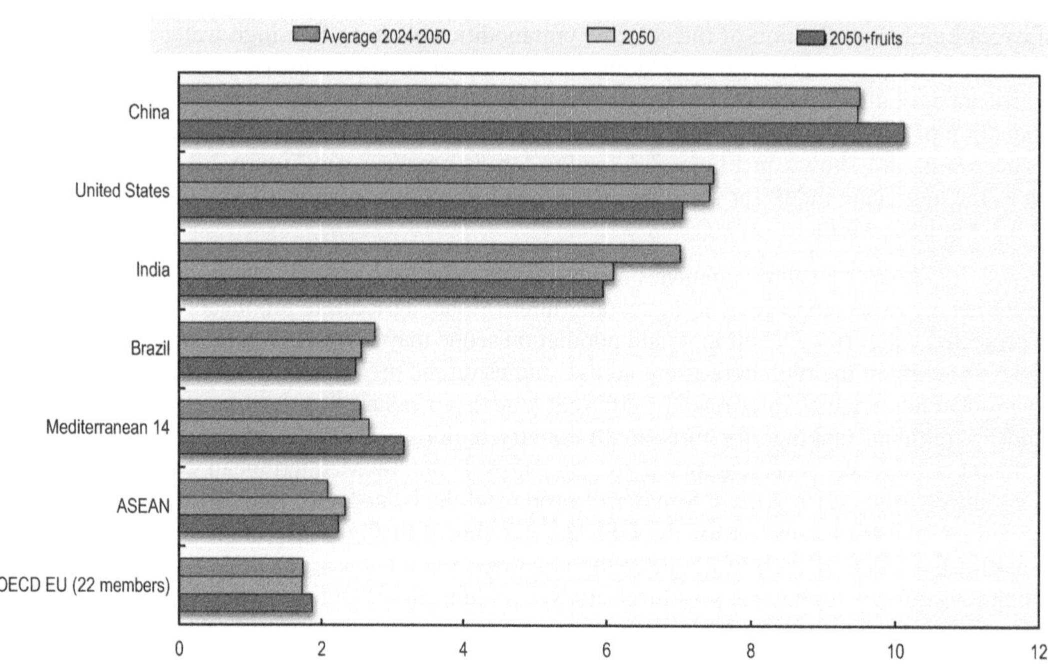

Note: Countries of the Mediterranean 14 region are: Morocco, Algeria, Tunisia, Libya, Egypt, Jordan, Israel, Lebanon, Syria, Turkey, Greece, Italy, France and Spain. For ASEAN, the represented countries are Myanmar, Cambodia, Lao PDR, Viet Nam, Thailand, Malaysia, the Philippines and Indonesia

Source: Compilations using AgLink-Cosimo and IMPACT projections and results from the review of 64 publications.

Figure 2.9. Future water risk hotspots in global agriculture production (2024-2050 average)

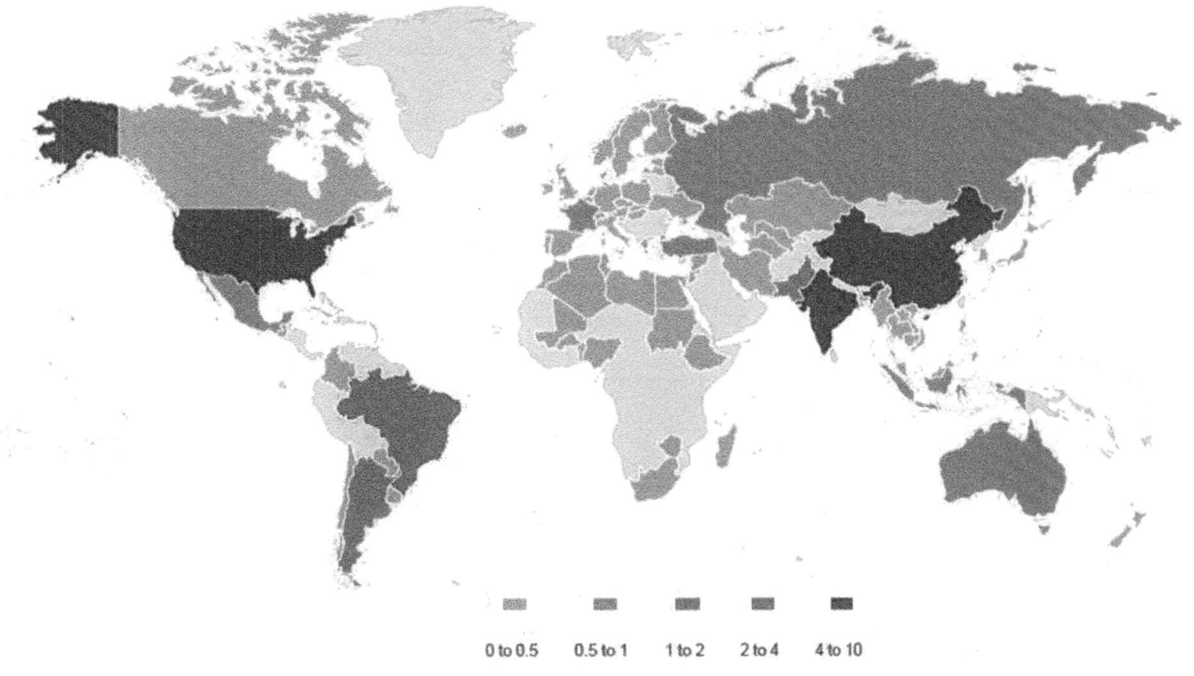

Note: The index can be interpreted as the expected share of overall global production of the key agriculture commodities likely to face high water risks without adaptation action in each of the 77 largest agriculture-producing countries.

Source: Review of 64 publications, AgLink-Cosimo and IMPACT simulations.

These rankings should not mask the importance of differences in water risks and agriculture activities. Figures 2.A2.2 and 2.A2.3 in Annex 2.A2 give a decomposition of the same indicators by the eight commodities plus fruit, using the IMPACT projections for 2050. It should be noted that these national level measurements may be inaccurate for large countries, where specific agriculture activities do not coincide with water risks. With this important limitation in mind, Table 2.3 shows the three leading countries for each commodity, together with high risk supranational regions and their hypothetical rankings. China, India and the United States are the leading countries in terms of production at risk under almost all categories based on these indicators, followed by Brazil and Indonesia as major producers of certain products. At the supranational level, Indonesia contributes to the higher rank of ASEAN for rice and oilseed production. At the same time, the fruit sector in Mediterranean countries is expected to face major water risks, and OECD EU countries as whole face significant water risks in some sectors.

Table 2.3. Countries leading in agriculture future water risk indicators in 2050

	Country-based	Regions with high risks (theoretical rank)
Beef	1. China (7.7), 2. United States (6.7), 3. Brazil (3.8)	
Dairy	1. India (14.8), 2. China (6.3), 3. United States (5.8)	4. OECD EU members (2.9)
Coarse grains	1. United States (20.0), 2. China (12.5), 3 Brazil (2.3)	4. OECD EU members (1,6)
Rice	1. China (16.5), 2. India (11.8), 3. Indonesia (1.8)	3. ASEAN (6.4)
Wheat	1. China (8.5), 2. United States (5.0), 3. India (4.1)	5. OECD EU members (2.9)
Sugar	1. Brazil (7.2), 2. India (5.4), 3. China (4.3)	5. OECD EU members (2.5)
Cotton	1. China (14.9), 2. United States (10.9), 3. India (5.3)	
Oilseeds	1. United States (7.0 2. Indonesia (5.5), 3. China (5.0)	1.ASEAN (8.9)
Fruits	1. China (15.2) 2. India (4.8), 3. United States (4.0)	2. 14 Mediterranean countries (7.0)

Source: Author's own work, derived from the combined analysis of water risks and agriculture regions. See full results in Annex 2.A2.

These three countries also dominate in the different classes of water risks. The disaggregated indicators are shown for the three leading countries at risk in Table 2.4. While all countries are expected to be subject to the risk of water shortages, excess water is reported to be more of a problem in parts of China and India than in the United States.

Table 2.4. Share of categorical indices of future water risk in the three hotspot countries

	China	India	United States
Shortage	73%	58%	71%
Excess	65%	65%	30%
Variability	50%	0%	75%
Quality	33%	67%	67%

Source: Author's own work.

The combination of these two characterisations shows that different commodities will face different types of risks. Box 2.2 shows that coarse grains, cotton and fruits areas are found to be the most affected by water shortages.

Box 2.2. What type of water risks could affect specific agriculture commodities in 2050?

A broader question is what type of products may be most at risk. The indicators for future water risk for the 77 countries (and representing over 89% of global production of each commodity) can be used to derive commodity-based indicators for the entire market. The results (Figure 2.10) suggest that the areas most at risk are those that are projected to produce coarse grains, cotton, fruits, rice and dairy. Overall, these aggregate indicators suggest that 40% to 50% of global production of these commodities could face future water risks.

Figure 2.10. Aggregate future water risk indices (%) by commodity group in 2050

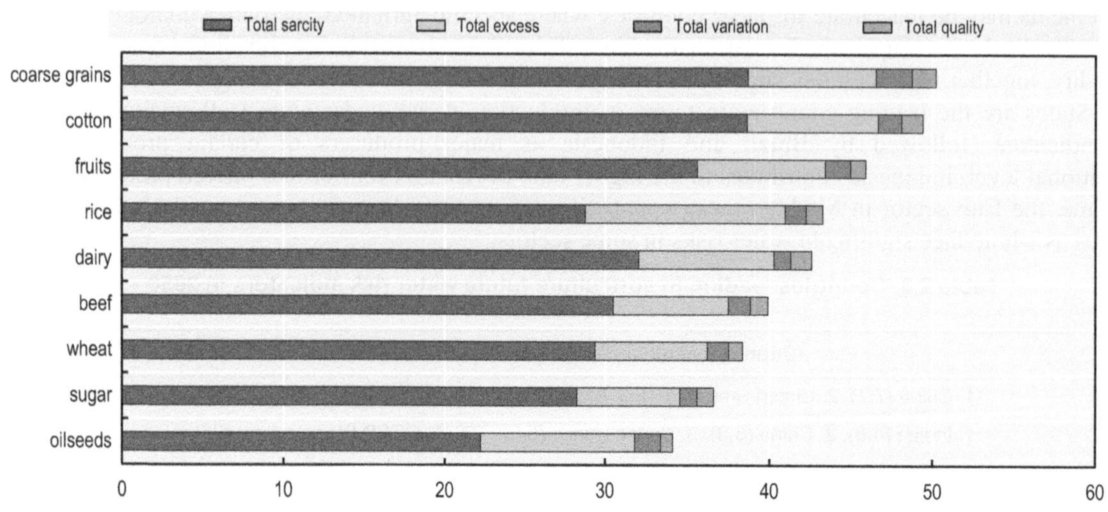

More specifically, Figure 2.11 shows the *relative* importance of the type of water risks in driving average future water risks by commodity, measured by computing standardised indices of changes. It shows that the risks of scarcity and extreme events (droughts) are especially important for countries that produce coarse grains, cotton and fruits (and relatively less oilseeds, sugar and wheat), while excessive water is projected to affect mostly rice and oilseed production areas (and less so sugar, beef, wheat or coarse grains). Water quality risks are projected to be more important in areas with coarse grains, cotton, and dairy, and lower in countries with oilseeds sugar and wheat.

Figure 2.11. Relative importance of the types of risks affecting commodity groups in 2050

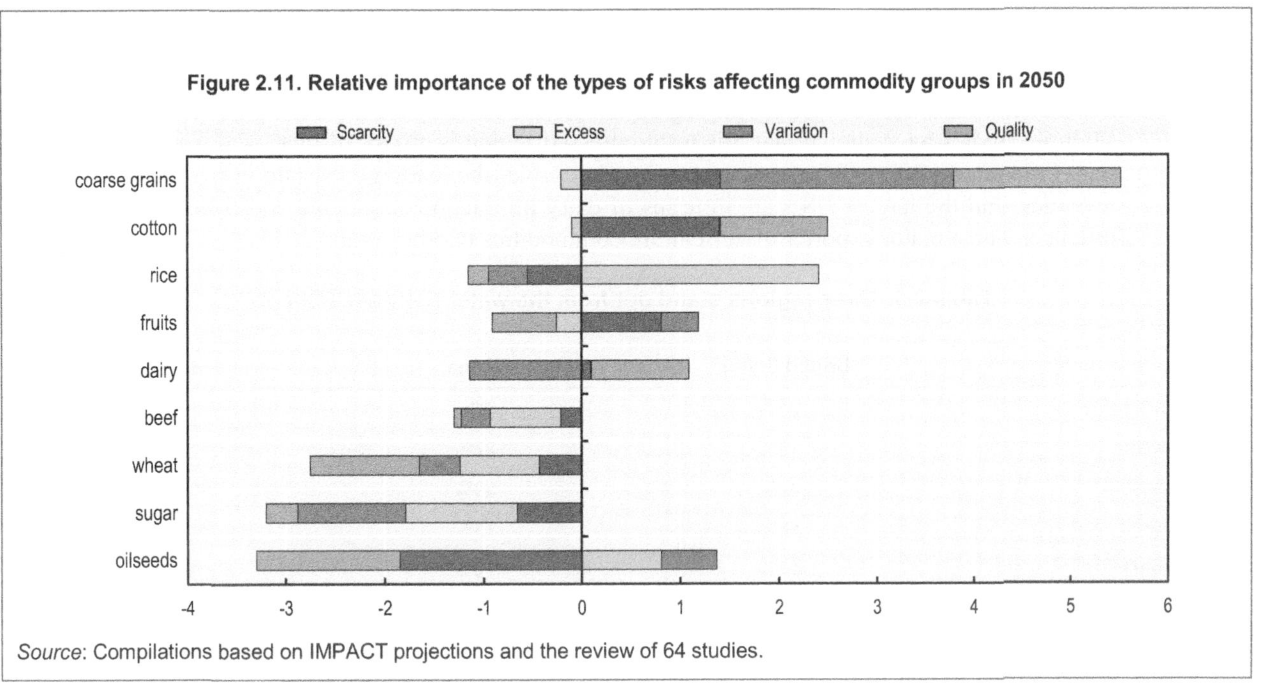

Source: Compilations based on IMPACT projections and the review of 64 studies.

China, India and the United States are enormous in size, but not all subnational regions are identical in agriculture production activities, nor subject to the same future water risks. Ideally, a full representation of subnational hotspots would require a complete global assessment of subnational water risks and commodity importance. Several reviewed studies did integrate water risks with agriculture, but they often focused on a single dimension of risks, and used specific scenarios and models that also face limitations.[12]

The following sections look at agriculture regions facing high water risks within each identified hotspot country. This second step considers national level risks, but nonetheless aims at targeting significant regional agriculture water risk, as the regions are selected based on the same criteria in countries that concentrate global agriculture production and water risks.

2.4. High agriculture water risks for agriculture in Northeast China, Northwest India and the Southwest United States

Identifying regions with high agriculture water risks in the three countries

The literature review undertaken can help single out regions that are most at risk in the three hotspot countries. Table 2.5 shows the frequency of reported water risks in specific regions. Combined with information about major agriculture production area, the following three regions are identified.[13]

- In China, most reported observations (48) locate prevalent water risks in the semi-arid northeast, which is the largest agriculture production region for cereals and cotton. It is also a largely populated and industrial economic region, with high levels of competition for water, and intense groundwater depletion.
- In India, several studies project high water risks throughout the country, with most observations (53%) pointing towards the cereal producing regions of the north and especially the northwest. and others (35% of the observations) reporting flooding or drought risks in the south-east, which has more diverse types of agriculture. Northwest India is known as the breadbasket of India and one of the world's hotspot for groundwater depletion due to intensive irrigation (e.g. Taylor et al., 2013).

Based on these criteria it is expected to be one of the most important agriculture regions facing water risks in India and worldwide.

- In the United States, the Southwest region is facing the most water risks, with 46 observations. Multiple studies have shown that this water stressed region is at risk of increased water constraints (e.g. Cook et al., 2015). The region includes California, the leading US state in overall agriculture revenues, and the largest US state in terms of dairy production, vegetable and fruits, a large cattle producer, and a major exporter of agriculture commodities (Cooley et al., 2016).

Table 2.5. Three regions stand out from the water risk assessment

	United States*	China	India
Central	23%	2%	9%
Northwest	19%	27%	**53%**
Northeast	9%	**59%**	39%
Southwest	**58%**	13%	32%
Southeast	15%	18%	35%

Note: *4% also identified Alaska and 11% Texas. Only observations with regional differences are accounted for. Current and future risks are pooled.
Source: Author's own derivations based on the reviewed literature.

The rest of this chapter will focus on these three key regions, analysing their agriculture and water risk specificities. While the present assessment cannot confirm their standing in comparison with subnational regions in other countries, collected literature-based evidence reviewed in this report does suggest their high concentration of water risks and agriculture production even at the global level. These three national hotspot regions also represent different agriculture structure and levels of development, providing a good basis to study more specifically how water risks can impact productive agriculture regions.[14]

Similar water trends across the three regions

The three regions face a number of similarities: a diverse agriculture production, increasing shortages of usable groundwater, and unstable surface water levels unstable given the pressure from other sectors.

Although characterised by extreme water scarcity (Figure 2.12), Northeast China[15] is a key area for agriculture production, industry development and population growth. National statistics show that in 2013, this region accounted for about 25% of wheat, corn and cotton, and 10% of rice and apple total production. With 8% of China's total water resources and one-third of China's population and GDP, competition for water resources is high (Liu and Speed, 2009). This incongruity has increased in recent decades due to declining water supply, lower precipitation, deteriorating water quality, and rising water demand (Hijoka et al., 2014; Box 2.3).

Figure 2.12. Northeast China concentrates agriculture and water scarcity

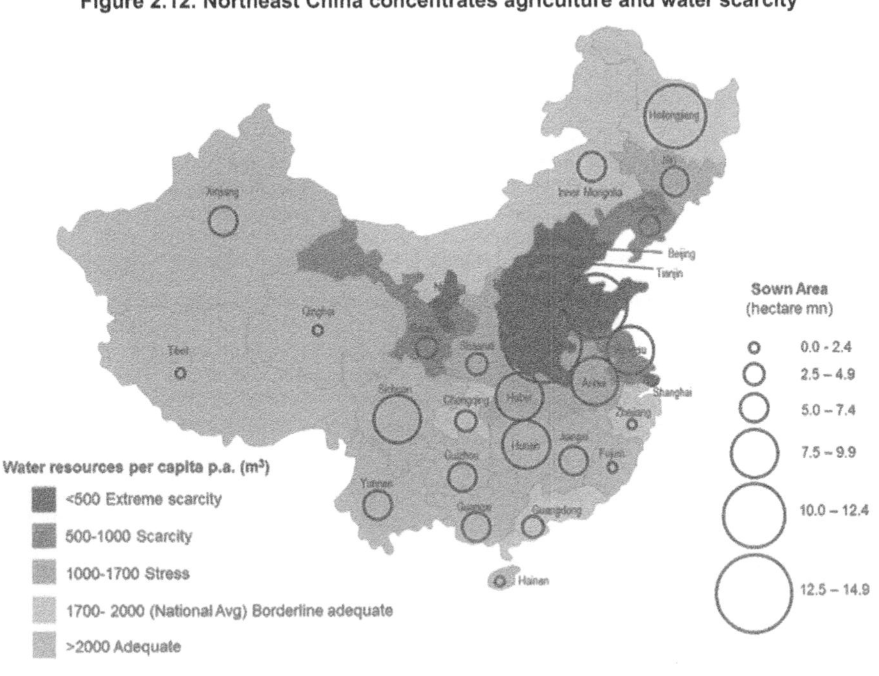

Source: Tan (2014). http://chinawaterrisk.org/resources/analysis-reviews/the-state-of-chinas-agriculture/.

Box 2.3. Factors that have increased water risks in Northeast China

The increasing frequency of more severe droughts, as well as the generally declining level of average precipitation from 1960 to the early 2000s, have considerably limited the *water supply* in Northeast China (Piao et al., 2010). Although the estimated level of water resources in the Northeast has stabilised (Figure 2.A2.5 in Annex 2.A2), the decrease in the groundwater level in the North China Plain — an agriculturally important sub-region of the Northeast — by about 1m per year over the last 20 years (Figure 2.13) is of critical concern (Foster and Garduño, 2014; Kendy et al., 2003; Chen, 2010, Giordano, 2009, Zhang and Diaz, 2014). This phenomenon, largely driven by intensive pumping for irrigation (Cao et al., 2013), has generated flow cut-offs from the Yellow River and its tributaries, and has contributed to the disappearance of 194 natural lakes and 40% of the region's waterways, in addition to allowing for coastal seawater intrusion and land subsidence in certain areas (Jiang, 2009; Moore, 2013; Sun et al., 2010; Wang and Jin, 2006; Wang et al., 2007).

Figure 2.13. Average levels of groundwater pumping and water tables in the North China Plain

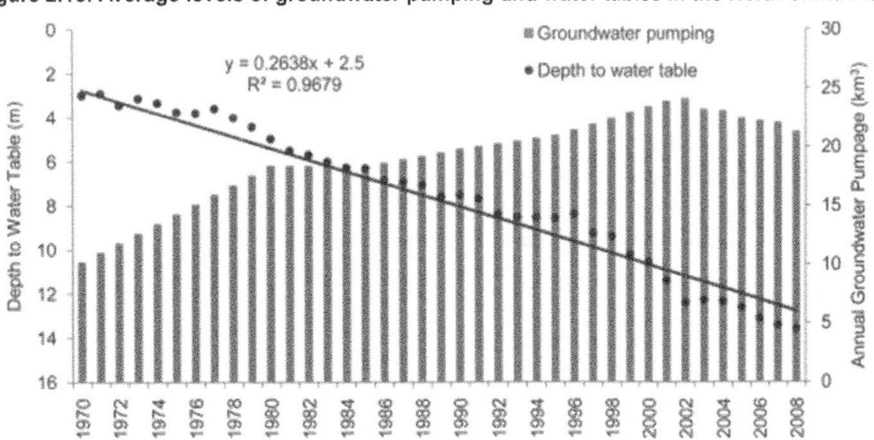

Note: Water tables were simulated based on best available information. Annual groundwater pumping before 1980 is estimated.

Source: Cao et al. (2013).

Water quality issues are also exacerbating water stress for agriculture in the Northeast. Water quality in the Yellow River declined steadily from 1985 to 2001 (Giordano, et al., 2004, see Figure 2.A2.6 in Annex 2.A2). Water in Classes IV and V – which is only suitable for industrial or agriculture use – increased from 4% in 1985 to 25% in 2001. One-third of the studied Yellow River water was deemed unfit even for agriculture in 2007 (Branigan, 2008). Although water quality of the Yellow river has reportedly improved in recent years, the quality of other rivers in the Northeast – such as the Huai and Liao – have declined (China Water Risk, 2015). Furthermore, 70% of the rural North China Plain's groundwater is too polluted at present for human use and could be harmful for agriculture (China Water Risk, 2016.)

Rising water demand is also a factor in water stress. In line with national trends, the agriculture sector is the largest water user in the Northeast (Figure 2.A2.5 in Annex 2.A2). In contrast to the 8% increase in agriculture use at the national level, agriculture use in the Northeast remained relatively steady from 2004 to 2014. During the same period, use by other sectors increased 20% in the Northeast, but only 13% at the national level. Increased production of relatively water-intensive crops such as corn – and higher transpiration rates due to excessive fertiliser use – have also reduced the water content of topsoil in Northern China (Liu et al., 2015).

Lastly, water stress in Northeast China has been compounded by *inefficient water usage*. The effective utilisation ratio of water (water effectively used over water withdrawn) in Chinese agriculture has improved from 44% in 2002 to 52% in 2013, but remains approximately 20 percentage points below the ratio in developed countries (Yu, 2016). Producing 1 kg of grain in China requires twice the water needed in developed countries (0.96 m^3) (Zhao et al., 2008). This is partly due to China's open channel irrigation system, which is highly susceptible to leakages. Water inefficiency may be particularly high in the Northeast due to low precipitation levels and a higher reliance on irrigation. Only a fraction of irrigated land in Northeast China is equipped with water-saving irrigation technologies (Huang et al., 2017).

Source: Branigan (2008); Cao et al. (2013); Chen (2010); China Water Risk (2015), Foster and Garduño (2004); Giordano (2009); Giordano et al. (2004); Huang et al. (2017); Kendy et al. (2003); Liu et al. (2015); Piao et al. (2010); Yu (2016); Zhang and Diaz (2014); Zhao et al. (2008).

The negative impact of water stress on agriculture production can already be seen for water-sensitive crops in drought years. For instance, corn production declined in the droughts of 1997, 1999, 2000, 2009 and 2014. Wheat production also declined in several years following droughts; sharp decreases can be found in 1998, 2000-2003; a slowdown also occurred in 2007 and 2009-2010.

Recent shifts in the Northeast's production basket do not appear to have reduced the region's water demands; overall production has increased, though its composition has evolved. Shares of corn and apple production have increased from 2004 to 2013 (Figure 2.A2.7 in Annex 2.A2). At the same time, production has shifted away – in relative terms – from cotton, rice, beans and tubers (and wheat to a certain extent). Taking into account the water requirements of these crops, water stress may have been reduced by the relative decline in cotton, rice and bean production and relative increase in apple production in recent years. However, it is unclear whether rising production of corn – a relatively water-intensive crop – may have offset this gain.

At the interface between the Indus and the Ganges river basins, Northwest India, here defined by the states of Punjab and Haryana, is the breadbasket of the country (Figure 2.14 left panel). These states belong to the Indo-Gangetic Belt, a fertile region with important groundwater and surface water resources, supplied by the snowmelt water of the Himalaya and annual monsoon rains. However, the climate is characterised by high inter-seasonal and inter-annual variability in precipitation, which makes them particularly exposed to drought (Punjab Department of Revenue, Rehabilitation and Disaster Management, 2014). Total water availability in Punjab is estimated at 39.5 km3 (32 Million acre-feet), while demand—largely borne by the agriculture sector—reaches 61.7km3 (50 Million acre-feet) (Punjab Department of Irrigation, 2008). This 38% deficit puts water reserves under pressure and causes groundwater depletion at a rapid pace. Haryana faces the same problem as Punjab: a rising gap between water demand and supply, which puts into question its capacity to sustain its large agriculture sector if water consumption does not become drastically more efficient.

Figure 2.14. A key grain production region under increasing groundwater stress

Food grain production 2010-2011 (tonnes/capita), Net groundwater availability for irrigation in 2025

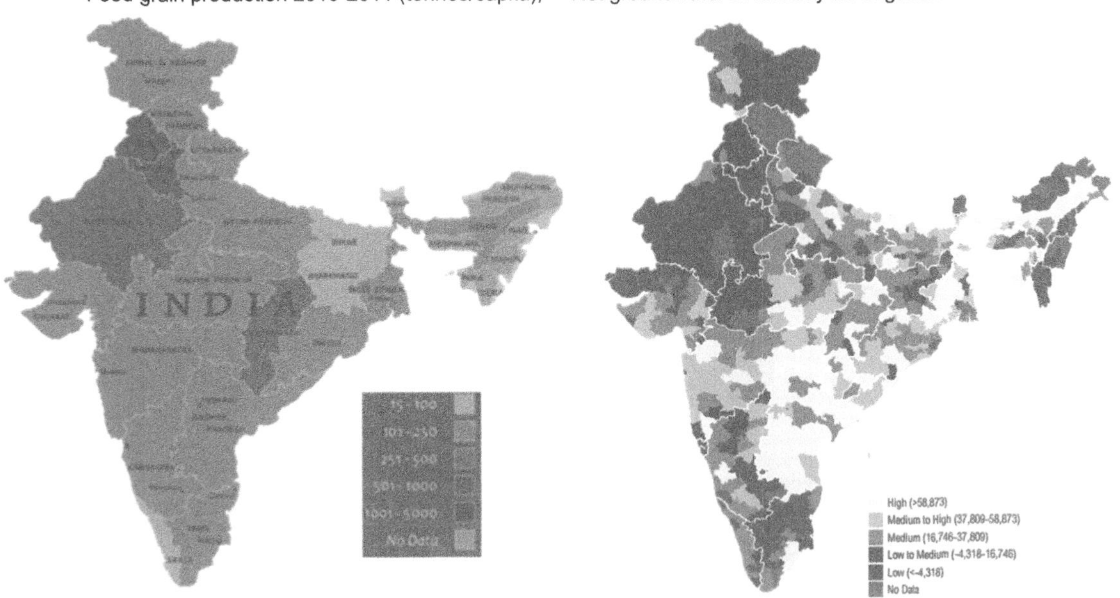

Source: Iyer (2013), https://www.scribd.com/document/202944804/Food-An-Atlas ; Shiao et al. (2015), http://www.wri.org/resources/maps/india-water-tool

Agriculture in this region depends on groundwater irrigation. To respond to a dry climate, the quasi totality of the cropped area in Punjab and Haryana is irrigated. Irrigation development and the use of high yielding seed varieties that are sensitive to drought have been encouraged since the Green Revolution (Murgai, 2001). Double cropping, first during the monsoon and then during the dry season, further increases the need for water input (Ministry of Statistics and Program Implementation, 2016a; Indian National Informatics Centre, 2016a). Groundwater irrigation expanded rapidly from the 1960s, pushed by affordable private tube well extraction systems—particularly suited to fragmented landholding (Banerji et al. 2010; World Bank, 2013b)[16]—, by government programmes encouraging intensive rice and wheat cultivation, and later by rural electricity subsidies (Box 2.4). It gradually replaced the use of a large but outdated canal network.[17] At present, respectively 53% and 71% of the sown area in Haryana and Punjab is under tube well irrigation. In comparison, only 21% of the sown lands are under tube well irrigation countrywide (Ministry of Statistics and Program Implementation, 2016b).

Box 2.4 Productive agriculture at the cost of depleting aquifers: The role of policies

From 1960 to 2009, the cropping intensity in Punjab rose from 126% to 190%, and the very water-intensive rice-wheat rotation covered the majority of the area (80% of cropped area), in part due to agriculture price policies. The area under paddy cultivation is about 2.8 million hectares in Punjab, while groundwater recharge capacity can only sustain 1.6 million hectares, leading to the continuous lowering of water table levels (CFAPPS, 2013). Farmers are incentivised to grow rice and wheat since food grain prices are guaranteed by the government, and their yields vary less than for other crops (Singh, J., 2013). Minimum support prices for these two crops also strongly limit potential crop diversification, necessary for a transition toward a less water-intensive agriculture (Garduño et al., 2011). Recent diversification efforts in Punjab are targeting shifts mainly to maize, cotton, sugarcane, and basmati rice, which need less water than Indica rice, the predominant variety (CFAPPS, 2013).

State subsidies for irrigation, in the form of energy subsidies, is a central factor explaining groundwater demand and overdraft in Punjab and Haryana. Whereas a falling water table level increases the amount of energy needed to lift water, this rising cost is not directly born by the farmers, who benefit from a free power supply (Badiani et al., 2012; CFAPPS, 2013). Therefore, there are few incentives to experiment less water-consuming cropping methods or to invest in water-saving technologies, even if growing cereals is increasingly energy-costly (Sharma et al., 2015) (Figure 2.15). In 2003, electricity subsidies to agriculture weighed 7.36% of Punjab state expenditures, which was more than the state budget allocated to health and education together (Birner et al., 2011). As of 2016, they still represented 6.8% of the State budget (Krar, 2016). In Haryana, where the farming sector consumes 40% of electrical power, the State decided to pursue and extend power subsidies to farmers. Thus, the allotted amount per cropped area rose by 42% between 2010 and 2014, in spite of heavy budget difficulties (Kumar et al., 2011; *The Times of India*, June 2014).

Figure 2.15. Energy efficiency and agriculture growth in Punjab

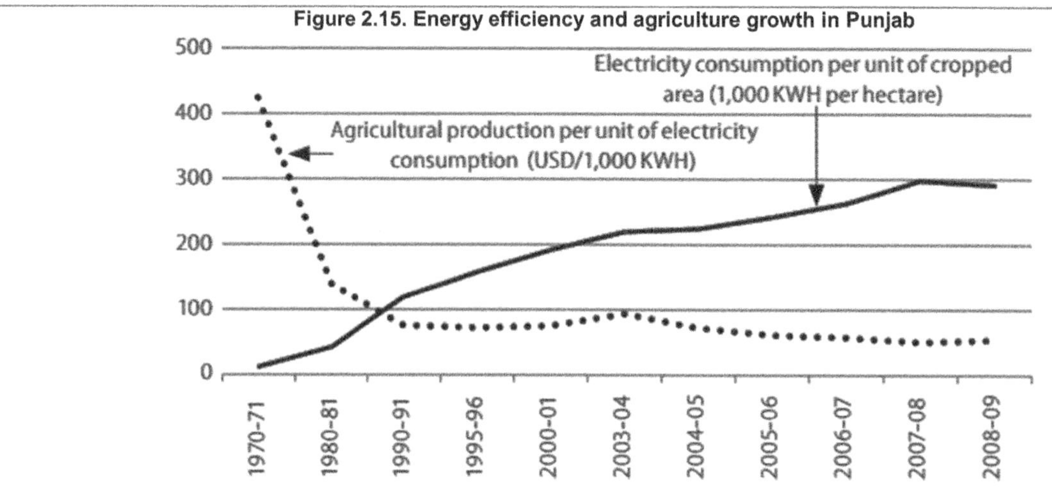

Source: Sarkar and Das (2014).

Power restrictions, instituted for energy control, have deferred the evolution of agriculture practices. Free energy supply for irrigation purposes is also made at the expense of the maintenance of the electricity network, which frequently collapses. As a result, farmers have installed automatic power switches to pump water as soon as power is available. This uncontrolled and unmetered irrigation system encourages excessive water extraction (Planning commission, 2009). In order to limit the use of water pumps, the State authorities determine pre-announced restricted hours during which free power will be available for irrigation. In Haryana, farms receive power six to ten hours a day, distributed in several phases (World Bank, 2001). In Punjab, power is available from four to eight hours a day (Sarkar and Das, 2014). These conditions are not adapted to drip irrigation, which requires a continuous power supply for at least eight hours (Punjab Department of Soil and Water Conservation, 2016). Therefore field flooding, which has very low water use efficiency, remains the most commonly used irrigation technique.

1. Cropping intensity can be defined as "the fraction of the cultivated area that is harvested. [It] may exceed 100 percent where more than one crop cycle is permitted each year on the same area." See www.fao.org/nr/water/aquastat/data/glossary/search.html?termId=7587&submitBtn=s&cls=yes

Source: Badiani et al. (2012); Birner et al. (2011); CFAPPS (2013); Garduño et al. (2011); Krar (2016); Kumar et al. (2011); Planning Commission (2009); Punjab Department of Soil and Water Conservation (2016); Sarkar and Das (2014); Singh (2013); World Bank (2001).

The rapid development of groundwater irrigation has broadly contributed to the productivity gains of Indian agriculture during the Green Revolution, but the continuous rise of tube well irrigation has led to an overexploitation of water reserves (Venkata and Burke, 2013) (Figure 2.14 right). In the beginning of the 2000s, groundwater consumption started overwhelming the recharge capacity in the States of Haryana and Punjab (Gandhi and Namboodiri, 2009). Ten years later, despite the growing awareness of central authorities, the tendency has dramatically intensified: in 2011, groundwater development stress[18] reached an average of 133% in Haryana, 172% in Punjab, and up to 416% in some districts (CGWB, 2014). In 2010 water level fell beyond a depth of 15m in 75% of Punjab, a significant increase from 2000 and 1980 when 14% and 0.6% of the area fell under this level (CFAPPS, 2013) (Figure 2.A2.10 in Annex 2.A2). As of 2016, 51% of the local administrative units called blocks[19] in Haryana and 75% of the blocks in Punjab are considered as over-exploited (CGWB, 2016). This phenomenon has been confirmed by satellite gravity data measurements that showed a rapid depletion of groundwater from 2002-2008 (Rodell, 2009). Groundwater depletion has also contributed to exacerbating water quality concerns (MacDonald et al., 2016).[20]

The Southwest United States[21] —defined here as Arizona, California, Colorado, Nevada, New Mexico, and Utah (six bottom left states on Figure 2.16)—is a rapidly growing important agriculture region that faces major water challenges. It has the fastest-growing population and one of the most economically important regions of the United States. It is also the nation's most arid region and increasingly prone to long-term droughts (Prein et al., 2016). The region is among the most productive agriculture regions in the world, generating 17% of national agriculture sales (BEA, 2015), and leading in US production and exports of many agriculture products. Agriculture in this region largely relies on irrigation and the sector contributes to an overuse of water resources, as observed by the rapid depletion of groundwater in the Central Valley of California (4.8 km^3 per year, Famiglietti et al., 2011) and the irregular flow of the Colorado River, which in the last 50 years has rarely reached the sea.

Recent trends show that agriculture has reduced its water use. While agriculture still accounts for about 70% of total freshwater withdrawal, water withdrawals declined from 1990 to 2010 by 18% across the region (Maupin et al., 2014). Between 2003 and 2013, total irrigated land declined by 6% in the Southwest, irrigated pastureland decreased by 11%, and rainfed pastureland increased by 25%. These changes may have been responses to the persistent drought conditions and the overall diminution of precipitations, as observed in Figure 2.16, or to long-term market trends.

Figure 2.16. Weather patterns that bring rains are increasingly rare in Southwestern United States

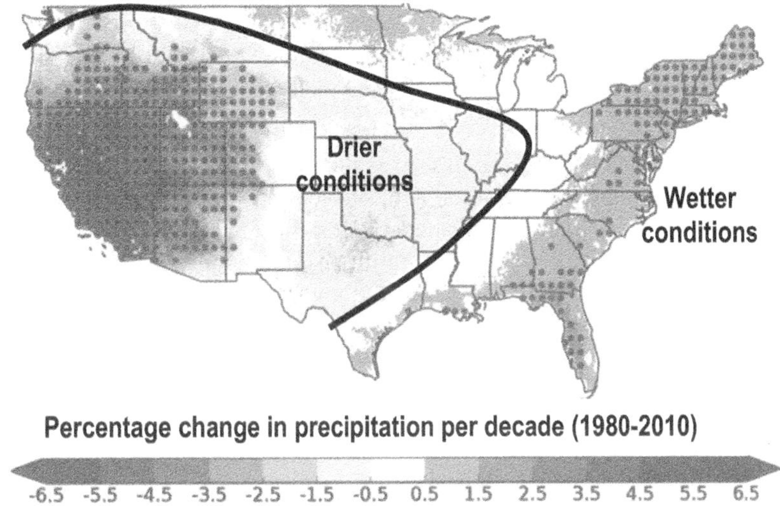

Note: This map depicts the portion of overall changes in precipitation across the United States that can be attributed to these changes in weather system frequency. The line delimits areas with negative change in precipitation (drier conditions) from those with positive precipitation change (wetter conditions). The grey dots represent areas where the results are statistically significant.
Source: Prein et al. (2016). https://www2.ucar.edu/atmosnews/news/19173/southwest-dries-wet-weather-systems-become-more-rare

Despite declines in water reserves, agriculture sales in the region increased by nearly 28% between 2005 and 2014, driven largely by California crop production, particularly fruits and nuts. The Southwest region doubled its agriculture exports over the same period. Much of this growth can be attributed to increased exports of meat and dairy products. The value of beef and veal exports, for example, increased nearly five-fold over the last decade, and the value of dairy exports tripled. Since 2005, the export value of crop products has increased by about 86 per cent.

These seemingly counter trends suggest that the agriculture sector has shown remarkable adaptability and resilience to droughts. Yet recent developments suggest this increase may not be sustainable. The 2011-16 drought in California encouraged the intensification of groundwater irrigation, leading to the accelerated depletion of the Central Valley aquifer as well as generating large negative external problems (Phillips et al., 2015).[22] The observed shift in farming from field crops to higher value permanent crops (fruits and nuts) may also have reduced the farming sector's adaptability to drought.

Growing water risks expected in the future

Although projections are subject to uncertainty, agriculture water stress in Northeast China is likely to worsen overall due to a combination of factors (see Annex 2.A2.3 for details).

- Climate change is projected to raise temperatures, leading to increased evaporation and to the melting of critical glaciers. The projected impact is mixed and uncertain on regional precipitation. The former suggests that short-term surface water scarcity relief could be accompanied by long-term, seemingly irreversible, water scarcity challenges (NARCC, 2007; Piao et al., 2010; Tao et al., 2003; Thomson et al., 2006; Zhang et al., 2007).

- Demand for water will continue to increase due to China's continued economic development, population growth and urbanisation dominating the overall water balance (2030 WRG, 2009). The largest supply-demand water deficit in 2030 (39%) is expected to be found in this region (Hai Basin, see 2030 WRG, 2012). Water demand in China's Northeast is predicted to increase across all sectors, with agriculture losing shares to other sectors (e.g. Table 2.A2.2 in Annex 2.A2).
- Despite a projected increase in recharge, groundwater in the North China Plain is projected to continue to be under pressure (Cao et al., 2013; Döll and Fiedler, 2008; FAO and WWC, 2015; Taylor et al., 2013).
- As industry expands, surface and groundwater quality is expected to further deteriorate due to rising pollution levels which will reduce the availability of usable water for agriculture (World Bank, 2013a; China Water Risk, 2016).

Northwest India is expected to face a higher water supply-demand gap, increased inter-seasonal rainwater variability, and potential deterioration of surface and groundwater quality (see details in Annex 2.A2.3). Water demand will increase, pushed by non-farming demand; in the next ten years, the region's population will increase by 5.8 million, and the urban population by 43% (National Commission on Population, 2006). By 2025, Punjab projects a 38% increase of domestic and industrial groundwater consumption, as compared to 2011 (CGWB, 2014). Climate change is expected to increase rainfall variability, leading to floods during the monsoon (Döll, 2002; IPCC, 2008), lower surface water supply and groundwater recharge and potential increase in irrigation water demand, exacerbating the pressure on aquifers (Bruinsma, 2003; Krishan et al., 2015; Mohinder, 2016; Taylor et al., 2013). Northwest India is projected to be one of the only Indian regions where groundwater levels are expected to decrease at a faster rate with climate change and irrigation requirements (Zaveri et al., 2016). The intensive exploitation of aquifers is expected to worsen groundwater quality, potentially reducing its availability for agriculture. Saline intrusion in groundwater, already expanding in Punjab, is expected to increase (Hill-Clarvis et al., 2016; Kim, 2013; Mahajan, et al., 2012), potentially affecting surface water (BGS, 2015). Growing industrialisation may also impact surface and groundwater usability.

The Southwest United States will also have to cope with more variable and uncertain water supplies due to the combination of lower and more volatile water supplies in critical river basins and projected higher demand. Climate change is projected to exacerbate the demand supply gap in the Colorado River Basin (USBR, 2012) (Box 2.5). Annual runoff in the northern part of the California Central Valley is projected to show little or no change, while some drying is projected in the southern portions of the region (USBR, 2014). Warmer temperatures during the winter months, however, cause precipitation to fall as rain rather than snow, increasing winter run-off and reducing spring run-off, which will reduce the ability of the state to store water in reservoirs (Ibid.).[23] Continued population growth may put additional pressure on the region's limited water resources; by 2030, an additional 23 million people are expected to live in the region bringing the area's population to 73 million, 48% higher than in 2000 (United States Census Bureau, 2005).

Box 2.5. Projecting water risks: The Colorado Basin

The Colorado River has played an important role in the development of the US Southwest and parts of northwestern Mexico, supplying water to about 40 million people in both countries, and wholly or partially irrigating more than 2 million ha inside and outside of the basin. A dense yet dynamic set of regulations, interstate compacts, agreements, contracts, judicial decisions, and an international treaty, known collectively as the "Law of the River", governs the allocation and use of water from the Colorado River water within the United States and from the United States to Mexico. Although massive dams along the river's main stream and many of its tributaries can store as much as four times the Colorado River's average annual flow, these have devastated ecosystems and driven several native fish species to the brink of extinction. These institutional and structural controls severely constrain the river's natural variability and significantly reduce the volume of water flowing to the border. In the past several decades, the Colorado has rarely had enough water to reach the sea.

Figure 2.17. Historical and projected Colorado River Basin water use and demand

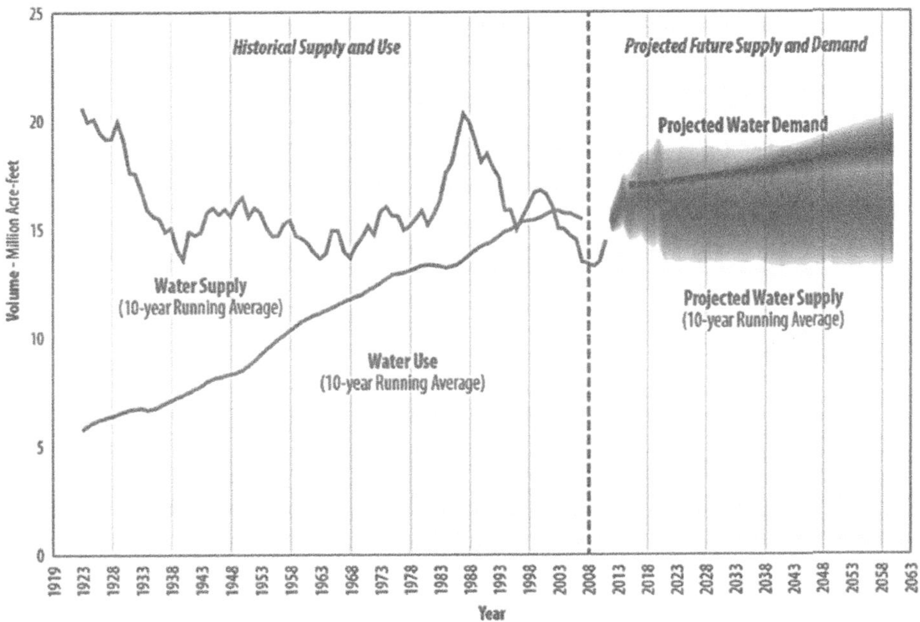

Note: Water use and demand include Mexico's allotment and losses, e.g. those due to reservoir evaporation, native vegetation, and operational inefficiencies.

Source: USBR (2012).

Beginning in 2010, the US Bureau of Reclamation partnered with Colorado River Basin stakeholders to evaluate future water supply and demand in the basin through the year 2060 and develop mitigation and adaptation strategies to address water supply and demand imbalances. The Colorado River Basin Study (USBR, 2012) was finalised in 2012 and included four future water supply scenarios to help capture the range of potential futures and reflect the uncertainty of projecting water supply in a highly variable system. In the 100-year historic record, mean annual runoff at Lee's Ferry – the traditional measuring point for the Colorado River - has been about 18 km3, with more than 80% of the runoff generated from about 15% of the Colorado River Basin at elevations exceeding 2 400 meters. The climate change scenario projects a general drying trend in the basin, with the notable exception of increased precipitation in the higher elevation, productive headwaters regions. With climate change, total runoff was projected to decline by 9.1%, with greater annual and monthly hydrologic variability in an already dynamic system, reducing predictability and reliability for water managers. The other water supply scenarios, based on the 100-year historic record and the much longer tree-ring record, projected that runoff would decrease by less than this amount. As shown in Figure 2.17, supply and demand imbalances already exist in the Colorado Basin and are projected to become more severe in the future.

Source: USBR (2012).

Notes

1. The hotspot approach as defined here includes both the hotspot identification exercise and the targeted response. Annex 2.A1 provides a methodological note with a discussion on water risk definitions.

2. Targeting water risks may also result in neglecting other types of risks for agriculture. The current approach does not claim to respond to all agriculture problems.

3. It should be noted that the approaches presented here focus mainly on the benefits of risk mitigation and not cost mitigation. In other words, targeting water risks for agriculture will not focus on whether low-cost solutions exist, but rather where the most critical risks may lie. This does not prevent considering cost-effective solutions to these problems.

4. For instance, a key objective of the OECD Council Recommendation on Water (OECD, 2016b) is "The application of pollution control measure as close to the source as possible".

5. If the studies are done at the global level, they may not always cover all countries thoroughly.

6. A secondary step is taken by identifying agriculture production regions facing water risks within each of the identified countries in section 2.4.

7. The overall methodology and list of countries are available in Annex 2.A2. It includes OECD country results.

8. Many countries in sub-Saharan Africa are not found to be expected to face higher water risk in the reviewed literature. As noted above, this may reflect the fact that many of the climate projections do not foresee increasing risks in these regions compared to others, and some see water risks actually diminishing, even if they start from a higher risk level.

9. If these three countries are large, the fact that other large or larger countries do not feature in this list, such as Brazil, Canada or the Russian Federation, but that small countries like Lebanon or Kuwait do, shows that the scale bias is not predominant with this method.

10. This asymmetry may be due to uncertainties of projections. One could expect less consistent estimates in the future than current measures, and the fact that there are fewer data points for current than future risks may create a bias.

11. This was done by aggregating their agriculture water risk indices for each region.

12. Researchers at the World Resource Institute have developed a mapping tool to indicate agriculture's exposure to water stress by agriculture commodity, but the analysis used recent or current data, without projecting future risks. http://wri.org/applications/maps/agriculturemap/#x=0.00&y=-0.00&l=2&v=home&d=rice

13. There was no sufficient data and publications to conduct a similar quantitative assessment of water risks and subnational agriculture projections in the three countries, but the review of literature conducted in each region (presented in following subsections) concurs with this selection.

14. Interestingly, the regions found to be most at risks are also significant agriculture regions. The link between the importance agriculture production and water risks can be partially explained by climatic and other non-water risk-related factors, including the quality of the land, rural development, etc.

15. For this analysis, data is drawn from ten provinces in Northeast China classified with extreme water scarcity: Beijing, Gansu, Hebei, Henan, Jiangsu, Liaoning, Ningxia, Shandong, Shanxi and Tianjin (Tan, 2014).The region includes four main rivers: the Huang (Yellow), Huai, Hai and Liao rivers. The Yellow River is the second longest river in China and its river basin connects to the Hai and Huai rivers.

16. A tubewell in Punjab irrigates 2.8 hectares of crops in average, but when considering small holdings (1 to 2 ha) one tubewell provides water to only 1.2 hectares (Indian National Informatics Centre, 2016b).

17. The two states are also in conflicts with regards to the sharing of river waters mostly for canal irrigation (Mangat, 2016).

18. Groundwater development stress is defined as the current annual rate of groundwater abstraction divided by the mean annual natural groundwater recharge.

19. Blocks are the assessment units used by the Central Groundwater Board to follow local water level trends. Punjab counts 138 blocks, and Haryana 108.

20. Within 60% of the aquifer under the Indo-Gangetic Basin, potable water is limited due to arsenic or excessive salinity (MacDonald et al., 2016).

21. This section is based on Cooley et al. (2016). It should be noted that the Southwest Region defined here excludes the Central and Southern High Plains Aquifer region, where agriculture also faces high water stresses and where water demand exceeds supply (OECD, 2015a).

22. California passed its first legislation of groundwater in 2014 (Box 4.3 in Chapter 4), but the implementation of the law is projected to take until 2042.

23. Under current reservoir operating criteria, "with earlier runoff and more precipitation occurring as rainfall, reservoirs may fill earlier and excess runoff may have to be released downstream to ensure adequate capacity for flood control purposes". (USBR, 2014: 4).

References

ADRC (Asian Disaster Reduction Centre) (2016), Global Identification number (GLIDE) database of international disasters, consulted December 2015, http://gs.adrc.asia/glide/public/search/search.jsp

Alavian, V. et al. (2009). "Water and Climate Change: Understanding the Risks and Making Climate-Smart Investment Decisions", the World Bank, Washington DC.

Alcamo, J. et al. (2007), "Future long-term changes in global water resources driven by socio-economic and climatic changes", *Hydrological Sciences Journal*, Vol. 52/2, pp. 247–275. doi:10.1623/hysj.52.2.247.

Amarasinghe, U.A. et al. (2008), "India's Water Supply and Demand from 2025-2050: Business-as-usual Scenario and Issues", International Water Management Institute (IWMI), Colombo, Sri Lanka.

Arnell, N. W. (2004), "Climate change and global water resources: SRES emissions and socio-economic scenarios", *Global Environmental Change*, Vol. 14, pp. 31–52. doi:10.1016/j.gloenvcha.2003.10.006.

Aulakh, M.S. et al. (2009), ''Water Pollution Related to Agricultural, Industrial, and Urban Activities, and its Effects on the Food Chain: Case Studies from Punjab", *Journal of New Seeds*, Vol. 2, pp. 112-137.

Badiani, R. et al. (2012) "Development and the Environment: The Implications of Agricultural Electricity Subsidies in India", *Journal of Environment and Development*, Vol. 21/2, pp.244–262.

Baettig, M.B., et al. (2007), "A climate change index", *Geophysical Research Letters*, Vol. 34, p. L01705.

Banerji, A. et al. (2010), "Social Contracts, Market and Efficiency: Groundwater Irrigation in North India", Working Paper No. 183, Centre for Development Economics, Delhi School of Economics.

Barnett, T.P. et al. (2005), "Potential impacts of warming climate on water-availability in snow-dominated regions", *Nature*, Vol. 438/17, pp.303-309.

Bates, B.C., et al. (eds.) (2008), "Climate Change and Water", Technical Paper of the Intergovernmental Panel on Climate Change, IPCC Secretariat, Geneva, 210 pp.

BEA (Bureau of Economic Analysis) (2015). "Regional Economic Accounts: SA45 Farm Income and Expenses", US Department of Commerce, Washington DC.

Bennett, B. (1998), "Rising salt, a test of tactics and techniques", *ECOS*, Vol. 96, pp. 10-13.

Bhupal, D.S. (2012), "Agricultural profile of Haryana", Agricultural Economics Research Center, University of Delhi, November 2012.

Bijlsma et al. (1996), "Coastal zones and small islands. In: Watson, R.T. et al. (eds.) *Climate Change 1995—Impacts, Adaptations and Mitigation of Climate Change: Scientific-Technical Analyses,* Cambridge University Press, Cambridge, UK, pp. 289–324.

Birner, R. et al. (2011), "The Political Economy of Agricultural Policy Reform in India - Fertilizers and Electricity for Irrigation", IFPRI Research Monograph, International Food Policy Research Institute, Washington DC.

Brackenridge, G.R. (2016) "Global Active Archive of Large Flood Events", Dartmouth Flood Observatory, University of Colorado, Boulder, CO. Consulted January 2016. http://floodobservatory.colorado.edu/Archives/index.html

Branigan, T. (2008), "One-third of China's Yellow river 'unfit for drinking or agriculture'", *The Guardian*, London. http://www.theguardian.com/environment/2008/nov/25/water-china

Brauman, K. A. et al. (2015), "Water depletion : An improved metric for incorporating seasonal and dry-year water scarcity into water risk assessments", *Elemental: Science of the Anthropocene*, Vol. 4/83, pp.1–12. http://doi.org/10.12952/journal.elementa.000083

BGS (British Geological Survey) (2015), "Groundwater resources in the Indo-Gangetic Basin - Resilience to climate change and abstraction", British Geological Survey Open Report, Natural Environment Research Council, 31 July 2015.

Bruinsma, J. (2003), "World agriculture: Towards 2015/2030 – An FAO Perspective", FAO, Earthscan Publications Ltd, London.

Buckle, S. and F. Mactavish (2013), "The changing water cycle", Grantham Briefing Note 5, University College of London, London.

Buckley, C (2015), "The Findings of China's Climate Change Report", *The New York Times*, November 30, 2015.

Cao, G. et al. (2013), "Use of flow modelling to assess sustainability of groundwater resources in the North China Plain", *Water Resource Research*, Vol. 49, pp. 159–175

CCA (Council of Canadian Academies) (2013), "Water and Agriculture in Canada: Towards Sustainable Management of Water Resources", CCA, Ottawa, Canada. http://www.scienceadvice.ca/en/assessments/completed/water-agri.aspx

CCAFS (CGIAR Research Program on Climate Change, Agriculture and Food Security) and CIMMYT (International Maize and Wheat Improvement Centre) (2014), "Climate-Smart Villages in Haryana, India", Research Program on Climate Change, Agriculture and Food Security, New Delhi. https://cgspace.cgiar.org/rest/bitstreams/34314/retrieve

CFAPPS (Committee for Formulation of Agriculture Policy for Punjab State) (2013), "Agriculture Policy for Punjab", Report submitted to Government of Punjab, March 2013.

CGWB (Central Ground Water Board) (2016), State Profile, Government of India, Ministry of water resources, river development and Ganga rejuvenation, Faridabad, updated in March 2016, http://cgwb.gov.in/gw_profiles/st_AP.htm

CGWB (2015), "Groundwater Year Book of Haryana State (2014-2015)", Government of India, Ministry of water resources, river development and Ganga rejuvenation, Chandigarh.

CGWB (2014), "Dynamic Ground Water Resources of India, as on 31st March 2011", Government of India, Ministry of water resources, river development and Ganga rejuvenation, Faridabad.

CGWB (2013), "Master Plan for artificial recharge to groundwater in India", Government of India, Ministry of water resources, New Delhi.

CGWB (2006), "Dynamic Ground Water Resources of India, as on March 2004", Government of India, Ministry of water resources, Faridabad.

Chavas, D.R. et al. (2009), "Long-term climate change impacts on agricultural productivity in eastern China", *Agricultural and Forest Meteorology*, Vol. 149, pp. 1118–1128.

Chen J. Y. (2010), "Holistic assessment of groundwater resources and regional environmental problems in the North China Plain", *Environmental and Earth Science*, Vol. 61, pp.1037–1047.

Chen, S. et al. (2016), "Impacts of climate change on agriculture: Evidence from China", *Journal of Environmental Economics and Management*, Vol. 76, pp.105-124. http://dx.doi.org/10.1016/j.jeem.2015.01.005

China Water Risk (2016), Website, Last accessed February 2016. http://chinawaterrisk.org/

China Water Risk (2015), "2014 State of Environment Report Review", China Water Risk, Hong Kong. http://chinawaterrisk.org/resources/analysis-reviews/2014-state-of-environment-report-review/

Ciscar, J.C. et al. (2014), "Climate Impacts in Europe", The JRC PESETA II Project, Joint Research Centre (JRC) Scientific and Policy Reports, EUR 26586EN, European Commission, Seville, Spain.

Cook, B.I., et al. (2015), "Unprecedented 21st century drought risk in the American Southwest and Central Plains", *Science Advance,* Vol. 1, e1400082.

Cooley, H. et al. (2016), "Water Risk Hotspots for Agriculture: The case of the Southwest United States", OECD Food, Agriculture and Fisheries Paper N. 96, OECD Publishing, Paris.

Cosgrove, C. E., and W.J. Cosgrove (2011), "The Dynamics of Global Water Futures Driving Forces 2011 – 2050", UNESCO, Paris, France.

CRED (Center for Research on the Epidemiology of Disasters) (2016), "EM-DAT Glossary", EM-DAT, the international disasters database, Université Catholique de Louvain, Belgium http://www.emdat.be/Glossary (accessed 7 January 2016).

De Sherbinin, A. (2013), "Climate Change Hotspots Mapping : What Have We Learned ?" *Climatic Change*, Vol. 123/1, pp. 23-37.

Ding et al. (2007), "Detection, Causes and Projection of Climate Change over China: An Overview of Recent Progress", *Advances in Atmospheric Sciences*, Vol. 24, No. 6, pp. 954–971.

Döll, P. (2009), "Vulnerability to the impact of climate change on renewable groundwater resources", *Environmental Research Letters*, Vol.4/3. DOI:10.1088/1748-9326/4/3/035006

Döll, P. (2002), "Impact of climate change and variability on irrigation requirements: a global perspective", *Climatic change*, Vol. 54, pp.269-293.

Döll, P. and Fiedler, K. (2008), "Global-scale modeling of groundwater recharge", *Hydrology and Earth System Sciences*, Vol. 12, pp. 863–885.

Döll, P. et al. (2015), "Integrating risks of climate change into water management", *Hydrological Sciences Journal*, Vol. 60/1, pp.4-13. DOI: 10.1080/02626667.2014.967250

Elliott, J. et al. (2014), "Constraints and potentials of future irrigation water availability on agricultural production under climate change", *Proceedings of the National Academy of Sciences*, Vol. 111/9, pp. 3239–3244.

Famiglietti, J.S. et al. (2011). Satellites Measure Recent Rates of Groundwater Depletion in California's Central Valley, *Geophysical Research Letters*, Vol. 38, p. L03403.

FAO (Food and Agriculture Organization) (2015a), CLIMPAG Website, available at: http://www.fao.org/nr/climpag/hotspots/index_en.asp

FAO and WWC (World Water Council) (2015), Towards a Water and Food Secure Future: Critical Perspectives for Policy-makers, White Paper, FAO and WWC, Rome, and Marseilles. http://www.worldwatercouncil.org/fileadmin/world_water_council/documents/publications/forum_documents/FAO_WWC_Water_and_Food_WhitePaper.pdf

Fischer, G. et al. (2007), "Climate change impacts on irrigation water requirements: Effects of mitigation, 1990–2080", *Technological Forecasting & Social Change*, Vol. 74, pp.1083–1107.

Foster S. and Garduño H. (2004), China: towards sustainable groundwater resource use for irrigated agriculture on the North China Plain, Case Profile Collect, World Bank, Washington, DC.

Fraser, E.D.G., et al. (2013), "Vulnerability hotspots: Integrating socio-economic and hydrological models to identify where cereal production may decline in the future due to climate change induced drought", *Agricultural and Forest Meteorology*, Vol. 170, pp. 195-205.

Frenken, K. and V. Gillet (2012), "Irrigation water requirement and water withdrawal by country", Aquastat., FAO, Rome Available from: http://www.fao.org/nr/water/aquastat/main/index.stm

Fuhrer, J., et al. (2013), "Water Demand in Swiss Agriculture – Sustainable Adaptive Options for Land and Water Management to Mitigate Impacts of Climate Change", ART-Schriftenreihe 19, Agroscope, Swiss Confederation, Bern. https://www.agroscope.admin.ch/agroscope/en/home/topics/plant-production/field-crops/kulturarten/cereales/publikationen-sortenlisten-sortenpruefung/_jcr_content/par/columncontrols/items/0/column/externalcontent.external.exturl.pdf/aHR0cHM6Ly9pcmEuYWdyb3Njb3BlLmNoL2VuLVVTL0VpbnplbH/B1Ymxpa2F0aW9uL0Rvd25sb2FkRXh0ZXJuP2VpbnplbHB1Ymxp/a2F0aW9uSWQ9MzMyMjc=.pdf

Fung, F. et al. (2011), "Water availability in +2◦C and +4◦C worlds", *Philosophical Transactions of the Royal Society A*, Vol. 369, pp.99–116. doi:10.1098/rsta.2010.0293

Gandhi, V.P and Namboodiri, N.V. (2009), "Groundwater Irrigation in India: Gains, Costs and Risks", Indian Institute of Management, Ahmedabad.

Garduño, H. et al. (2011), "India, Groundwater Governance case study", Water Papers, World Bank, Washington DC.

Gassert, F. et al. (2013), Aqueduct Global Maps 2.0, The World Resources Institute, Washington DC.

Gerten, D. et al. (2011), "Global Water Availability and Requirements for Future Food Production", *Water and Global Change (WATCH)*, October, pp. 885–899. doi:10.1175/2011JHM1328.1

Giordano, M. (2009), "Global groundwater? Issues and solutions" *Annual Review of Environment and Resources*, Vol. 34/1, pp.153–178. doi:10.1146/annurev.environ.030308.100251

Giordano, M. et al.(2004), "Water Management in the Yellow River Basin: Background, Current Critical Issues and Future Research Needs", in IWMI Comprehensive assessment of water management in agriculture, International Water Management Institute, Colombo, Sri Lanka.

Giorgi, F. (2006), "Climate change hot-spots", *Geophysical Research Letters*, Vol. 33, p. L08707, doi:10.1029/2006GL025734.

GIST Advisory (2013), "Trends and patterns of food production in Punjab: Bio-physical aspects", GIST Advisory, Mumbai India.

Gleeson, T. et al. (2012), "Water balance of global aquifers revealed by groundwater footprint", *Nature*, Vol. 488, pp. 197-200.

Gosain, A.K. et al. (2001), "Climate change impact assessment of water resources of India", *Current Science*, Vol. 101/ 3, pp.356-371.

Haddeland, I. et al. (2014), "Global water resources affected by human interventions and climate change", *Proceedings of the National Academy of Sciences*, 111/9,pp. 3521–3256. doi:10.1073/pnas.1222475110

Hanasaki, N. et al. (2008), "An integrated model for the assessment of global water resources – Part 2: Applications and assessments", *Hydrology and Earth System Sciences*, Vol. 12, pp. 1027-1037.

Hare, W. L. et al. (2011), "Climate hotspots : key vulnerable regions, climate change and limits to warming", *Regional Environmental Change*, Vol. 11, pp. S1–S13. doi:10.1007/s10113-010-0195-4

Hatton, T. et al. (2011), "The Australian State of the Environment 2011", Report for the Australian Parliament, Canberra. https://www.environment.gov.au/science/soe/2011

Hejazi, M. I. et al. (2014), "Integrated assessment of global water scarcity over the 21st century under multiple climate change mitigation policies", *Hydrology and Earth System Sciences*, Vol. 18, pp. 2859–2883. doi:10.5194/hess-18-2859-2014

Hijioka, Y. et al. (2014), "Asia" in *Climate Change 2014: Impacts, Adaptation, and Vulnerability*, *Part B: Regional Aspects*, Contribution of Working Group II to the Fifth Assessment Report of the Intergovernmental Panel on Climate Change, Cambridge University Press, Cambridge, United Kingdom and New York, NY, USA, pp. 1327-1370.

Hill-Clarvis, M. et al. (2016), "Indian Insurance and Sustainable Development", Report for the British High Commission in New Delhi and the Federation of Indian Chambers of Commerce and Industry, Earth Security Group. http://ficci.in/spdocument/20722/FICCI-ESG-INSURANCE-REPORT.pdf

Hirabayashi, Y., et al. (2013), "Global flood risk under climate change", *Nature Climate Change*, Vol. 3, pp. 816–821. doi:10.1038/nclimate 1911

HLPE (2015), "Water for food security and nutrition", a report by the High Level Panel of Experts on Food Security and Nutrition of the Committee on World Food Security, Rome.

Huang, Q. et al. (2017), "Do water saving technologies save water? Empirical evidence from North China", *Journal of Environmental Economics and Management*, Vol. 82, pp.1-16.

IEG (Independent Evaluation Group) (2011), "Adapting to Climate Change: Assessing World Bank Group Experience" Phase III of the World Bank Group and Climate Change, IEG, Washington DC.

IFPRI and Veolia (2015), "The murky future of global water quality", White Paper, IFPRI and Veolia USA, Washington, DC.

Iglesias, A. and L. Garrote (2015), "Adaptation strategies for agricultural water management under climate change in Europe", *Agricultural Water Management*, Vol. 155, pp. 113–124.

Ignaciuk, A. (2015), "Adapting Agriculture to Climate Change: A Role for Public Policies", OECD Food, Agriculture and Fisheries Papers, No. 85, OECD Publishing, Paris. DOI: http://dx.doi.org/10.1787/5js08hwvfnr4-en

Ignaciuk, A. and D. Mason-D'Croz (2014), "Modelling Adaptation to Climate Change in Agriculture", OECD Food, Agriculture and Fisheries Papers, No. 70, OECD Publishing, Paris. DOI: http://dx.doi.org/10.1787/5jxrclljnbxq-en

Ignaciuk, A. et al. (2015), "Better drip than flood: reaping the benefits of efficient irrigation", *EuroChoices* Vol. 14/2, pp. 26-32.

Indian National Informatics Centre (2016a), "Irrigated area cropped once and more than once" (2011-2012), State Tables, in Input Survey Database, Agriculture Census Division, New Delhi. http://inputsurvey.dacnet.nic.in/statetables.aspx

Indian National Informatics Centre (2016b), ''District-wise Distribution of Characteristics of Holdings by Size Class/Size Group" (2010-2011), State Tables, in Agricultural Census Database, Agriculture Census Division, New Delhi. http://agcensus.dacnet.nic.in/StateSizeClass.aspx

IPCC (Intergovernmental Panel on Climate Change) (2012), "Managing the Risks of Extreme Events and Disasters to Advance Climate Change Adaptation", in [Field, C.B. et al. (eds.), *A Special Report of Working Groups I and II of the Intergovernmental Panel on Climate Change* Cambridge University Press, Cambridge, UK, and New York, US, 582 pp.

IPCC (2008), "Analysing regional aspects of climate change and water resources" in Climate Change and Water, Technical Paper VI, IPCC Secretariat, Geneva.

Iyer, K. (2013), "The Grain Necessities", in Jensen, D. et al. (eds), *Food: An Atlas*, map. https://www.scribd.com/document/202944804/Food-An-Atlas

Jiménez Cisneros, B.E., et al. (2014), "Freshwater resources". In Field, C.B. et al. (eds.), Climate Change 2014: Impacts, Adaptation, and Vulnerability: Part A: Global and Sectoral Aspects, Contribution of Working Group II to the Fifth Assessment Report of the Intergovernmental Panel on Climate Change, Cambridge University Press, Cambridge, UK and New York, pp. 229-269.

Kendy, E. et al. (2003), "Policies drain the North China Plain: agricultural policy and groundwater depletion in Luangcheng County, 1949–2000", Research Report 71, International Water Management. Institute (IWMI), Colombo, Sri Lanka.

Kiguchi, M., et al. (2015), "Re-evaluation of future water stress due to socio- economic and climate factors under a warming climate", *Hydrological Sciences Journal*, Vol. 60/1, pp. 14-29.

Kim, C-G. (2013), "The Impact of Climate Change on the Agricultural Sector: Implications of the Agro-Industry for Low Carbon, Green Growth Strategy and Roadmap for the East Asian Region", background policy paper for the East Asia Low Carbon Green Growth Roadmap project, Korea International Cooperation Agency.

Kleinen, T., and G. Petschel-Held (2007), "Integrated assessment of changes in flooding probabilities due to climate change", *Climatic Change*, Vol. 82, pp.283-312.

Krar, P. (21 July 2016), "Three firms submit bids for validating energy consumed by agricultural pump sets in Punjab", The Economic Times, India. http://economictimes.indiatimes.com/news/economy/agriculture/three-firms-submit-bids-for-validating-energy-consumed-by-agricultural-pump-sets-in-punjab/articleshow/53323392.cms

Krishan, G. et al. (2015), "Rainfall Trend Analysis of Punjab, India using Statistical Non-Parametric Test", *Current World Environment* Vol. 10/3, pp.792-800.

Kumar, M.D. et al. (2011), "Inducing the shift from flat-rate or free agricultural power to metered supply: Implications for groundwater depletion and power sector viability in India", *Journal of Hydrology*, Vol. 409, pp. 382–394.

Lapworth, D.J. et al. (2014), "Intensive Groundwater Exploitation in the Punjab – an Evaluation of Resource and Quality Trends", British Geological Survey, Open Report OR/14/068.

Lau, W.K. and K-M. Kim (2012), "The 2010 Pakistan Flood and Russian Heat Wave: Teleconnection of Hydrometeorological Extremes", *Journal of Hydrometeorology*, Vol. 13, pp. 392–403. http://dx.doi.org/10.1175/JHM-D-11-016.1

Liu, B., and R. Speed (2009), "Water Resources Management in the People's Republic of China", *International Journal of Water Resources Development*, Vol.25, pp.193-208.

Liu, J. et al. (2014), "International trade buffers the impact of future irrigation shortfalls", *Global Environmental Change*, Vol. 29, pp. 22–31. doi:10.1016/j.gloenvcha.2014.07.010

Liu, Y. et al. (2015), Agriculture intensifies soil moisture decline in Northern China, *Scientific Reports*, Vol. 5, Article number: 1126. DOI:10.1038/srep11261.

Luck, M. et al. (2015), "Aqueduct Water Stress Projections: Decadal projections of water supply and demand using CMIP5 GCMs", The World Resources Institute, Washington DC.

Lui, B. (2006) "China's Agricultural Water Policy Reforms", in OECD, *Water and Agriculture: sustainability, markets and policies*, OECD Publishing, Paris.

Luo, T., et al. (2015). Aqueduct Projected Water Stress Country Rankings, Washington D.C. http://www.wri.org/publication/aqueduct-projected-water-stress-country-rankings

MacDonald, A. M. et al. (2016), "Groundwater quality and depletion in the Indo-Gangetic Basin mapped from in situ observations", *Nature Geoscience*, Vol. 9, pp. 762-766.

Mahajan, G. et al. (2012), "Scope for enhancing and sustaining rice productivity in Punjab (Food bowl of India)", *African Journal of Agricultural Research,* Vol. 7/42, pp. 5611-5620.

Mangat, H.S. (2016), "Water war between Punjab and Haryana", *Economic and Political Weekly*, Vol. LI, N. 50.

Margat, J. and J. van der Gun (2013), Groundwater around the World: A Geographic Synopsis, CRC Press/Balkema, Taylor and Francis, London.

Maupin, M. et al. (2014),"Estimated Use of Water in the United States in 2010", US Geological Survey Circular 1405, USGS, Washington DC.

Mekonnen, M. M. and A. Y. Hoekstra (2016), "Four billion people facing severe water scarcity", *Science Advances*, Vol. 2/2, e1500323, DOI: 10.1126/sciadv.1500323

Mendelsohn, R. et al. (2012), "The impact of climate change on global tropical cyclone damage", *Nature Climate Change,* Vol. 2/3, pp. 205–209. doi:10.1038/nclimate1357

Milly, P. C. D. et al. (2005), "Global pattern of trends in streamflow and water availability in a changing climate", *Nature*, Vol. 438, pp. 347–50. doi:10.1038/nature04312

Ministry of Statistics and Program Implementation (2016a), "Pattern Of Land Utilisation, All India And State wise" (2011-2012), Agriculture Chapter in Statistical Year Book, India 2015 (database), Government of India, New Delhi. http://mospi.nic.in/Mospi_New/upload/SYB2015/ch8.html

Ministry of Statistics and Program Implementation (2016b),"Net area under irrigation by source" (2011-2012), Irrigation Chapter in Statistical Year Book, India 2015 (database), Government of India, New Delhi. http://mospi.nic.in/Mospi_New/upload/SYB2015/ch12.html

Ministry of Statistics and Program Implementation (2016c), "Average Yield Of Principal Crops, All India And State wise." and "Production Of Principal Crops, All India And State wise." (2011-2012), Agriculture Chapter in Statistical Year Book, India 2015 (database), Government of India, New Delhi. http://mospi.nic.in/Mospi_New/upload/SYB2015/ch8.html

Mohinder, G. (2016), "Direct Delivery of Power Subsidy to Agriculture", Presentation at the South Asia Groundwater Forum, Jaipur, 1 June 2016.

Murgai, R. (2001), "The Green Revolution and the productivity paradox: evidence from the Indian Punjab", *Agricultural Economics,* Vol. 25, pp. 199–209.

Murray, S. J. et al. (2012), "Future global water resources with respect to climate change and water withdrawals as estimated by a dynamic global vegetation model", *Journal of Hydrology*, Vol. 448-449, pp. 14–29. doi:10.1016/j.jhydrol.2012.02.044

NARCC (National Assessment Report on Climate Change)(2007), China's national assessment report on climate change, *Science Press*, Beijing.

National Bureau of Statistics of China (2016), Official statistic database, Beijing. (accessed January 2016).

National Commission on Population (2006), "Population projections for India and States 2001-2026", Office of the Registrar General & Census Commissioner, New Delhi, May 2006.

Nicholls, R.J. and A. Cazenave (2010), "Sea-Level Rise and Its Impact on Coastal Zones", *Science*, Vol. 328, pp. 1517–1521.

Nicholls, R.J. et al. (1999), "Increasing flood risk and wetland losses due to global sea-level rise: regional and global analyses", *Global Environmental Change*, Vol. 9, pp. S69-S87.

Nicholls, R.J., and R. S. J. Tol (2006), "Impacts and responses to sea-level rise: a global analysis of the SRES scenarios over the twenty-first century", *Philosophical Transactions of the Royal Society A*, Vol. 364/1841, DOI: 10.1098/rsta.2006.1754.

OECD (2016a), *Mitigating Droughts and Floods in Agriculture: Policy Lessons and Approaches*, OECD Studies on Water, OECD Publishing, Paris. http://dx.doi.org/10.1787/9789264246744-en

OECD (2016b), "OECD Council Recommendation on Water" [C(2016)174/FINAL], OECD, Paris. https://www.oecd.org/environment/resources/Council-Recommendation-on-water.pdf

OECD (2015a), *Drying wells, rising stakes: Towards sustainable agricultural groundwater use*, OECD Studies on Water, OECD Publishing, Paris. DOI: http://dx.doi/10.1787/978926423870-en

OECD (2015b), *Environment at a Glance 2015: OECD Indicators*, OECD Publishing, Paris. DOI: http://dx.doi.org/10.1787/9789264235199-en

OECD. (2015c), *The Economic Consequences of Climate Change*, OECD Publishing, Paris. DOI: http://dx.doi.org/10.1787/9789264235410-en

OECD (2014a), *Climate Change, Water and Agriculture: Towards Resilient Systems*, OECD Studies on Water, OECD Publishing, Paris. http://dx.doi.org/10.1787/9789264209138-en

OECD (2014b), "Global irrigation water demand projections to 2050: An analysis of convergences and divergences" [ENV/EPOC/WPBWE(2012)2/FINAL], OECD, Paris.

OECD (2013), *Water Security for Better Lives*, OECD Studies on Water, OECD Publishing, Paris. DOI: http://dx.doi.org/10.1787/9789264202405-en

OECD (2012), *Water Quality and Agriculture: Meeting the Policy Challenge*, OECD Studies on Water, OECD Publishing, Paris. http://dx.doi.org/10.1787/9789264168060-en

OECD (2010), *Sustainable Management of Water Resources in Agriculture*, OECD Studies on Water, OECD Publishing, Paris. http://dx.doi.org/10.1787/9789264083578-en

OECD (2009), *Managing Risk in Agriculture: A Holistic Approach*, OECD Publishing, Paris. http://dx.doi.org/10.1787/9789264075313-en

OECD/FAO (2015), OECD-FAO Agricultural Outlook 2015, OECD Publishing, Paris. DOI: http://dx.doi.org/10.1787/agr_outlook-2015-en

Pandey, S. (2011), "Report on impact assessment of notification in Gurgaon town and adjoining industrial areas of Gurgaon district, Haryana", Central Ground Water Board, North Western Region, Chandigarh, December 2011.

Parish, E.S., et al. (2012), "Estimating future global per capita water availability based on changes in climate and population", *Computers & Geosciences,* Vol. 42, pp 79-86.

PSCST (Punjab State Council for Science and Technology) (2014), "Punjab State Action Plan on Climate Change", Department of Science Technology and Environment, Government of Punjab, February 2014.

Pfister, S. and P. Bayer (2013), "Monthly water stress: spatially and temporally explicit consumptive water footprint of global crop production", *Journal of Cleaner Production*, Vol. 30, pp. 1-11.

Phillips, S.P. et al. (2015), "Sustainable groundwater management in California" U.S. Geological Survey Fact Sheet 2015-3084 (ver. 2.2, February 2016), USGS, Washington DC. http://dx.doi.org/10.3133/fs20153084

Piao, S. et al. (2010), "The impacts of climate change on water resources and agriculture in China", *Nature*, Vol. 467, pp.43-51.

Piontek, F., et al. (2013), "Multisectoral climate impacts in a warming world", *Proceedings of the National Academy of Sciences*, Vol. 111/9, pp. 3233-3238. doi:10.1073/pnas.1222471110.

Planning Commission (2009), "Report of the Task Force on Irrigation", Government of India, New Delhi.

PRC (People's Republic of China) (2007), "China's National Assessment Report on Climate Change", Science Publishing House, Beijing.

Prein, A.F. et al. (2016), "Running dry: The U.S. Southwest's drift into a drier climate state", *Geophysical Research Letters,* Vol. 43, pp.1272-1279. DOI: 10.1002/2015GL066727.

Punjab Department of irrigation (2008), "State Water Policy - 2008", public draft, Government of Punjab, Chandigarh.

Punjab Department of Soil and Water Conservation (2016), "Status and scope for drip and micro irrigation in Punjab", Government of Punjab, Chandigarh. http://dswcpunjab.gov.in/contents/micro_irrigation.htm (accessed August 2016)

Punjab Department of Revenue, Rehabilitation and Disaster Management (2014), "Climate of Punjab", Government of Punjab, http://punjabrevenue.nic.in/for%20website/Climate%20of%20Punjab.htm (accessed August 2016).

Punjab Directorate of Agriculture (2016), "Introduction to Punjab", Government of Punjab, Chandigarh, http://agripb.gov.in/home.php?page=intpu (accessed August 2016).

Rang, A. et al. (2014), "Status paper on rice in Punjab", Rice Knowledge Management Portal, Punjab Government, Chandigarh, www.rkmp.co.in/sites/default/files/ris/rice-state-wise/Status%20Paper%20on%20Rice%20in%20Punjab.pdf (accessed August 2016).

Reager, J. T. et al. (2016), "A decade of sea level rise slowed by climate-driven hydrology", *Science*, Vol. 351/6274, pp. 699–704.

Ringler, C. et al. (2011), *Sustaining growth via water productivity: Outlook to 2030/2050*, IFPRI and Veolia, Washington DC.

Robinson, S., et al. (2015) "The international model for policy analysis of agricultural commodities and trade (IMPACT); Model description for version 3". IFPRI Discussion Paper No. 1483, International Food Policy Research Institute, Washington, DC, http://ebrary.ifpri.org/cdm/ref/collection/p15738coll2/id/129825.

Rockström, J. and L. Karlberg (2009), "Zooming in on the global hotspots of rainfed agriculture in water-constrained environments", in S. P. Wani, et al. (eds.), *Rainfed agriculture: unlocking the potential*, CABI International.

Rockström, J. et al. (2010), "Managing water in rainfed agriculture—The need for a paradigm shift", *Agricultural Water Management,* Vol. 97/4, pp. 543–550, http://dx.doi.org/10.1016/j.agwat.2009.09.009.

Rockström, J. et al. (2009), "Future water availability for global food production: The potential of green water for increasing resilience to global change", *Water Resources Research*, Vol. 45, pp. 1–16. http://dx.doi.org/10.1029/2007WR006767.

Rodell, M. et al. (2009), "Satellite-based estimates of groundwater depletion in India", *Nature*, Vol. 460/7258, pp. 999-1002.

Rowley, R. J. et al. (2007), "Risk of Rising Sea Level to Population and Land Area", *Eos, Transactions American Geophysical Union* Vol. 88/9, pp. 105–107.

Sadoff, C. et al. (2015), "Securing Water, Sustaining Growth", Report of the GWP/OECD Task Force on Water Security and Sustainable Growth, Oxford University, Oxford, https://www.water.ox.ac.uk/wp-content/uploads/2015/04/SCHOOL-OF-GEOGRAPHY-SECURING-WATER-SUSTAINING-GROWTH-DOWNLOADABLE.pdf.

Sarkar, A. and A. Das, A. (2014), "Groundwater Irrigation-Electricity-Crop Diversification Nexus in Punjab: Trends, Turning Points, and Policy Initiatives", *Economic & Political Weekly*, vol xlix/ 52, pp. 64-73.

Schaldach, R. et al. (2012), "Current and future irrigation water requirements in pan-Europe: An integrated analysis of socio-economic and climate scenarios", *Global and Planetary Change*, Vol 94-95, pp. 33-45.

Scheffran, J. and A. Battaglini (2011), "Climate and conflicts: the security risks of global warming", *Regional Environmental Change,* Vol. 11(Suppl. 1), pp. S27-S39. http://dx.doi.org/10.1007/s10113-010-0175-8.

Schewe, J. et al. (2013), "Multimodel assessment of water scarcity under climate change", *Proceedings of the National Academy of Sciences*, Vol. 111/ 9, pp. 3245-3250, http://dx.doi.org/10.1073/pnas.1222460110.

Schlosser et al. (2014), "The future of global water stress: An integrated assessment", Report N. 254, MIT Joint Program on the Science and Policy of Global Change, MIT, Cambridge, MA. http://globalchange.mit.edu/files/document/MITJPSPGC_Rpt254.pdf

Shen, Y. et al. (2014), "Projection of future world water resources under SRES scenarios : an integrated assessment", *Hydrological Sciences Journal*, Vol. 59/10, pp. 1775–1793.

Shen, Y. et al. (2010), "Projection of future world water resources under SRES scenarios: Water withdrawal", *Hydrological Sciences Journal*, Vol. 53/1, 11–33, http://dx.doi.org/10.1623/hysj.53.1.11.

Shiao, T. et al. (2015), "India water tool", Technical Note, World Resources Institute, February 2015.

Shrivastava, P. and Kumar, R. (2014), "Soil salinity: A serious environmental issue and plant growth promoting bacteria as one of the tools for its alleviation", *Saudi Journal of Biological Sciences*, Vol. 22, pp. 123–131.

Siebert, S. et al. (2010), "Groundwater use for irrigation – A global inventory", *Hydrology and Earth Sciences*, Vol. 14, pp. 1863-1880.

Singh, I. and K. Singh Bhangoo (2013), "Irrigation system in Indian Punjab", MPRA Paper No. 50270, University of Munich, Munich, Germany.

Singh, J. (2013), "Depleting water resources of Indian Punjab agriculture and policy options - A lesson for high potential areas", *Global Journal of Science Frontier Research*, Vol. 13/4.

Singh, R.B. (2001), "Impact of land-use change on groundwater in the Punjab-Haryana plains, India", in *Impact of Human Activity on Groundwater Dynamics,* Proceedings of a symposium held during the Sixth IAHS Scientific Assembly at Maastricht, The Netherlands, IAHS Publication No. 269.

Tan, D. (2014), "The state of China's agriculture", China Water Risk, Hong Kong, China, http://chinawaterrisk.org/resources/analysis-reviews/the-state-of-chinas-agriculture/.

Tao, F. et al. (2009), "Modelling the impacts of weather and climate variability on crop productivity over a large area: A new superensemble-based probabilistic projection", *Agricultural and Forest Meteorology*, Vol. 149/8, pp.1266-1278.

Tao, F. et al. (2003), "Future climate change, the agricultural water cycle, and agricultural production in China", *Agriculture, Ecosystems and Environment*, Vol. 95, pp. 203–215.

Taylor, R. G. et al. (2013), "Ground water and climate change", *Nature Climate Change*, Vol. 3, pp.322-329.

Thom, et al. (2001), "Australia: State of the Environment 2001", Independent Report to the Commonwealth Minister for the Environment and Heritage, CSIRO, Canberra, https://soe.environment.gov.au/sites/g/files/net806/f/soe2001.pdf?v=1487243878.

Thomson, A.M. et al. (2006), "Climate change impacts on agriculture and soil carbon sequestration potential in the Huang-Hai Plain of China", Agriculture, *Ecosystems and Environment*, Vol. 114, pp. 195–209.

UNDESA (United Nations Department of Economic and Social Affairs) (2015), "World population prospects: The 2015 revision", United Nations, Department of Economic and Social Affairs, Population Division, New York, NY.

UNESCO (United Nations Education, Science and Cultural Organisation) (2010), "Climate change and adaptation for water resources in Yellow River Basin", China, IHP VII Technical Document in Hydrology UNESCO Office in Beijing, 2010.

USBR (United States Bureau of Reclamation) (2014), "Climate impact assessment: Sacramento and San Joaquin Basins", Prepared for Reclamation by CH2M Hill, USBR, Washington DC. www.usbr.gov/watersmart/wcra/docs/ssjbia/ssjbia.pdf

USBR (2012), "Colorado River Basin water supply and demand study", USBR, Washington DC. www.usbr.gov/lc/region/programs/crbstudy/finalreport/Study%20Report/CRBS_Study_Report_FINAL.pdf.

United States Census Bureau (2005), "Interim state population projections", USCB, Washington DC.

van Drecht, G. et al. (2009), "Global nitrogen and phosphate in urban wastewater for the period 1970 to 2050", *Global Biochemical Cycles*, Vol. 23, pp. 1-19.

van Puijenbroek, P.J.T.M. et al., (2014), "Global implementation of two shared socioeconomic pathways for future sanitation and wastewater flows", *Water Science and Technology*, Vol. 71, No. 2, pp. 227-233.

Venkata, S. and J. Burke (2013), "Smallholders and sustainable wells", Food and Agriculture Organisation, Rome.

Villholth, K.G. et al. (2014), "Groundwater in global irrigated food production – The role of depleting aquifers", International Water Management Institute (IWMI), Colombo, Sri Lanka.

Vörösmarty, C. J. et al. (2000), "Global Water Resources: Vulnerability from climate change and population growth", *Science*, Vol. 289, p.284.

Vrba, J., and A. Lipponen (eds.) (2007), "Groundwater resources sustainability indicators", United Nations Educational, Scientific and Cultural Organization (UNESCO)- International Hydrogeological Program, UNESCO, Paris.

Wada, Y. et al. (2013), "Multimodel projections and uncertainties of irrigation water demand under climate change", *Geophysical Research Letters*, Vol. 40, pp. 4626–4632, http://dx.doi.org/10.1002/grl.50686.

Wada, Y. et al. (2012), "Nonsustainable groundwater sustaining irrigation: A global assessment", *Water Resources Research*, Vol. 48 (November), p.W00L06.

Wang, J. (2017), "Agricultural groundwater use in China: Challenges, solutions and outlook", Presentation given at the Global Forum for Food and Agriculture (GFFA), Berlin, January 2017. https://www.slideshare.net/ifpri/agriculture-and-groundwater-feeding-billions-from-the-ground-up

Wang, J. et al. (2014), "Overview of impacts of climate change and adaptation in China's agriculture", *Journal of Integrative Agriculture*, Vol. 13/1, pp. 1-17.

World Bank (2013a), "China 2030: Building a modern, harmonious, and creative society", World Bank and the Development Research Center of the State Council, People's Republic of China, Washington, DC

World Bank (2013b), "Turn down the heat - Climate extremes, regional impacts, and the case for resilience", International Bank for Reconstruction and Development, The World Bank, Washington DC.

World Bank (2001), "India Power Supply to Agriculture", Report No. 22171-IN, 15 June 2001, The World Bank, Washington DC.

WWF (Word Wildlife Fund) (2016), "Overall water risk, food producers (including Tobacco)", Map of the Water Filter, WWF and KFW–DEG, WWF Germany, Berlin.

Xie, J. et al. (2008), "Addressing China's water scarcity: Recommendations for selected water resource management issues", World Bank, Washington DC, http://documents.worldbank.org/curated/en/2009/01/10170878/addressing-chinas-water-scarcity-recommendations-selected-water-resource-management-issues.

Xiong, W. et al. (2010), "Climate change, water availability and future cereal production in China", *Agriculture, Ecosystems and Environment*, Vol. 135, pp. 58-69.

Young, A. (1989), *Agroforestry for Soil Conservation*, CABI, Wallingford, UK.

Yu, W. (2016), "Agricultural and Agri-Environment Policy and Sustainable Agricultural Development in China", Background Report for the OECD Secretariat, March 2016.

Zaveri, E. et al. (2016), "Invisible water, visible impact: groundwater use and Indian agriculture under climate change", *Environmental Research Letters*, Vol. 11, Article 084005, http:dx.doi.org/10.1088/1748-9326/11/8/084005.

Zhang, L. et al. (2007), "Potential effects of climate change on runoff in the Yellow River Basin of China", *Transactions of the American Society of Agricultural and Biological Engineers*, Vol. 50, pp. 911–918.

Zhang, C. and L. Diaz, L. (2014), "A multi-regional input – output analysis of domestic virtual water trade and provincial water footprint in China", *Ecological Economics*, Vol. 100, pp.159–172, http://dx.doi.org/10.1016/j.ecolecon.2014.02.006.

Zhao, J., et al. (2008), "Opportunities and challenges of sustainable agricultural development in China", *Philosophical Transactions of the Royal Society B*, Vol. 363, pp. 893–904.

2030 WRG (2030 Water Resources Group) (2012), "The Water Resources Group: Background, Impact and the Way Forward", International Financial Centre, Word Bank Group, Washington DC.

2030 WRG (2009), "Charting our Water Future", International Financial Centre, Word Bank Group, Washington DC.

Annex 2.A1

Methodological note on water risks hotspot determination and examples of OECD applications

2.A1.1 Defining and measuring water risks

There are multiple definitions of risks (e.g. OECD, 2009). Some associate it with uncertainty around an event, others separate the two notions. For instance, in the context of natural disasters, risks are the "expected losses of lives, persons injured, property damaged and economic activity disrupted) due to a particular hazard for a given area and reference period" (CRED, 2016). A more general definition would be the "exposure to uncertain unfavourable economic consequences" (OECD, 2009).[1] Here we will use the latter definition, which is slightly more specific than to say that a risk is a "consequence of an event or an activity with respect to something that human value" (OECD, 2013).

Assessing risk requires the identification of three factors (hazard, exposure, and in the context of climate change, vulnerability) that are rarely assessed in a systematic way. Hazard and exposure are key factors needed to estimate a risk. A hazard is "the potential occurrence of a natural or human-induced physical event that may cause" an impact (e.g. flood damages) (IPCC, 2012). The exposure is "the presence of people; livelihoods; environmental services and resources; infrastructure; or economic, social, or cultural assets in places that could be adversely affected", e.g. the presence of a large population in a coastal area (Ibid.). When considering a population of heterogeneous individuals, vulnerability also plays a significant role. Vulnerability corresponds to "the propensity or predisposition to be adversely affected" (Ibid.), such as availability of sufficient stock of usable groundwater under drought. Vulnerability can be reduced by building a resilience to risks, e.g. by making investments that reduce the future vulnerability of a system (allocating land to annual crops instead of perennial crops under increasing water stress).[2] Multiple reports on water risks only focus on partial assessments, for instance by assuming—implicitly or not—that one factor is varying (often exposure) while the others are not (identical hazards, uniform vulnerability).

Water risks can be defined broadly as water-related challenges that threaten the ability of a user to secure water. In other words, such risks encompass any barrier to water security. OECD (2013) defines water security as the absence of four types of risks:

- Risk of insufficient water to meet demand in both the short and long-run, including drought;
- Risk of excess water, including flood;
- Risk of water of inadequate quality for a given use;
- Risk of disruption of freshwater systems, when pressure exceeds their coping capacity (resilience).

This definition will be used as reference thereafter, acknowledging that the project will focus primarily on the three first categories of risks that have been most studied in the literature and more directly linked to agriculture productivity (shortage, excess, and inadequate quality). Table 2.A1.1 presents a matrix of these risks and their three dimensions. Agriculture is subject to multiple water phenomena that it does not control, but it is also a major water using and a significant polluting activity (OECD, 2010 and 2012). Both exogenous and endogenous water risks have to be considered to fully understand the challenges for agriculture. Both types of risks can and often do impact agriculture.

Table 2.A1.1. Decomposing the main agriculture water risks

	Hazard	Exposure	Vulnerability
Water shortage	Insufficient water for plant development, especially for water intensive crops, Lack of pasture, unsuitability of area for livestock.	Depends on climate, natural endowment of surface and groundwater, competing demands, and type of crop or livestock.	Depends on the presence or absence of irrigation, access to wells, storage of water, water efficiency measures, institutional and policy mechanisms.
Water excess	Excess water preventing agriculture, soil erosion, surface and groundwater flooding, livestock health issues.	Depends on the geography of the location (elevation, slopes, distance from water ways), and on the type of crop.	Depends on the infrastructure limiting damages, institutional and policy mechanisms to cope with excess water.
Water quality deterioration	Unusable water systems, irreversible groundwater or soil contamination, health effects on livestock, plants and humans.	Depends on local hydrogeology, distance to the sources of pollution, timing of pollution, and type of agricultural activity and the toxicity, life span, concentration and volume of pollutants.	Depends on access to treatment, or alternative sources of usable water, institutional and policy mechanisms.

Source: Author's own work.

In a dynamic setting, another distinction can be made between changes in the average or in the variability of water related variables. In the former case, hazards may be increasing in impact, but remain consistent with respect to long-term trends in exposure. For instance, increasingly lengthy droughts or continuous water quality deteriorations are increasing risks that can be feasible to anticipate. For changes in the variability of water risks, hazards and/or exposure may increase in variability, creating much more challenging predictions (storms, etc.).

There are multiple means to quantify water risks operating at different scales or for different purposes (e.g. see OECD, 2013). Some of the key indicators, both general and agriculture-related are indicated in Table 2.A1.2. Some specific examples are outlined in Table 2.A1.3 (see also Jimenez-Cisneros et al, 2014: 249). At the watershed, national or international level, most of the effort has focused on assessing the risk of shortages. Groundwater is often differentiated—multiple indicators have been proposed (e.g. OECD, 2015a; Vrba and Lipponen, 2007) but are often less used. Water quality risks are more assessed locally and typically more difficult to evaluate at a broader level.

On the quantitative side, a number of water stress indicators rely on alternative versions of the withdrawal-to-availability ratio, defined as the annual water withdrawal divided by annual water availability at the basin scale, W/Q, where the W is annual freshwater off-stream withdrawal for agriculture, industrial and domestic sectors, and Q is annual renewable freshwater resources. Usually, the extent of water stress is categorised as no-stress ($W/Q < 0.1$), low stress ($0.1 < W/Q < 0.2$), moderate stress ($0.2 < W/Q < 0.4$), and high stress ($W/Q > 0.4$). This index has the advantage of reflecting the integrated effects of both the pressure from human society (the demand side) and hydrological system (the supply side) (Shen et al., 2014). An illustration is provided in Figure 2.A1.1, with water stress estimates (%) in OECD countries as of 2009.

Table 2.A1.2. Selected indicators of water risks found in international literature

Risk	Indicators	Time horizon	Definition	Example of studies
Quantity	Water stress index	Current and future	Annual water withdrawal divided by annual water availability at the basin scale.	Gassert et al. (2013)
	Water supply-demand gaps	Current	Difference between current demand and supply based on accounting methods	2030 WRG (2009)
	Crop water footprint	Current	Monthly water stress indices are multiplied by irrigation crop requirements	Pfister and Bayer (2013)
	Runoff and demand projections	Future	Projected runoffs from climate circulation models are compared with water demand projections (withdrawals and consumption)	Luck et al. (2015)
	Irrigation water requirements	Current and future	Assessing the current and future requirement using climate models and crop requirements	Döll (2002); Schaldach et al., (2012)
	Irrigation reliability	Current and future	Ratio of supply over demand for irrigation computed with crop and climate models	Ignaciuk et al. (2015)
	Non-renewable groundwater irrigation	Current	Combining crop water demand with surface water availability and groundwater recharge assessments	Wada et al. (2012)
	Groundwater development stress (GDS)	Current	Ratio of groundwater use over renewable recharge.	Margat and van der Gun (2013)
	Groundwater footprint	Current	A measure of GDS is multiplied by the area of use and then divided by the aquifer area.	Gleeson et al. (2012)
	Groundwater dependent crops	Current and future	Accounting for crops irrigated by groundwater over depleting aquifers	Villholth et al. (2014)
	Distribution of water-stressed basins	Current and future	Distributions of basins with various extents of water stress as defined by: (a) withdrawal to availability ratio (W/Q) and per capita water availability (Q/c).	Shen et al. (2014)
	People in water stressed areas	Current	Assessing cultivated rainfed land with climate and aridity indices	Rockström et al. (2010)
	Flood damages	Current	Estimation of economic damages	EM-Dat (CRED, 2016)
Variability	Irrigation reliability	Future	Climate projections on agriculture are mixed with scenarios on irrigation using a global multi-market partial equilibrium model	Ignaciuk and Mason D'Croz (2014)
	Water security for agriculture	Future	Full water security modelled as a counterfactual with a global multi-market partial equilibrium model of agriculture	Sadoff et al. (2015)
	Drought frequency	Current and future	Consecutive dry days per year	IPCC (2012)
	Flood propensity	Current and future	Period to a major precipitation event (compared to those in a reference period)	IPCC (2012)
	Population at risk of floods	Future	Population potentially affected by the change in probability of a 50 year (or lower) flooding event under climate change	Kleinen and Petschel-Held (2007)
Quality	Emissions simulation	Future	Loadings are modelled based on past data and simulated emissions are projected using a partial equilibrium model of agriculture	IFPRI and Veolia (2015)
	Salinity risks	Current and future	Projections of constraints via satellite based land and surface and ground measurements	Bennett (1998)

Source: Authors' own work based on the reviewed literature.

Table 2.A1.3. Examples of methodologies used to assess water risks

Water quantity risk indicator	Study	Models used to calculate water demand and supply	Models used to calculate agriculture requirement	Future scenarios considered
Water supply, water demand, water stress, seasonal variability	Luck et al. (2015)	6 CMIP5 GCMs and projections of socioeconomic variables were derived from the Shared Socioeconomic Pathways (SSPs) hosted at IIASA	Irrigated area and irrigation efficiency using country-level regressions	Two climate scenarios (RCP4.5 and RCP8.5) and two shared socioeconomic pathways (SSP2 and SSP3)
Water Stress Index	Ringler et al. (2011)	International Model for Policy Analysis of Agricultural Commodities and Trade (IMPACT).		Four Economic Growth scenarios: (BAU), Low-Carbon, Grey, and Blue World
Short term (2020-2030) and long term (2071-2100) climate impacts	Ciscar et al. (2014)	Long term: 2080'S DSSAT, the models simulate daily phonological development and growth in response to environmental factors (such as climate including the effect of C02) and management (considering crop variety, N fertilisation, irrigation). The main output is "agricultural yield change". Short term: Compared to the baseline for the year 2000 on the CropSyst crop model.	BioMA-cropSyst, to assess crop yields (t/ha)	ENSEMBLES E1 scenario and Reference (BAU) simulation models
Water Stress Indicators	Vörösmarty, et al. (2000)	Water balance model (WBM) to calculate runoff	Irrigated land area and national use statistics	Climate Change data: Canadian Climate Center general circulation model (CGCM1) and the Hadley center circulation model HadCM2
The expected gap between 2030 demand figure and currently planned supply	2030 WRG (2009)	Water 2030 global water supply and demand model	Ag. Production based on IFRI IMPACT-water base case	Country specific base case scenarios for India, China, South Africa and Sao Paulo
Water management in rainfed agriculture	Röckstrom et al. (2010)	Link between climate and poverty is investigated. Thereafter, the number of people living in water-constrained agricultural areas is estimated. Based on this analysis, the global hotspots for rainfed agricultural areas in water-constrained environments are identified. Data on land use were derived from the Global Land Cover data set (GLC2000, 2003), in which the class 'cultivated and managed areas' was chosen to represent the total agricultural area. Second, a data set produced by the FAO (Food and Agriculture Organization, the United Nations) was used to represent irrigated agricultural land use. Water constraints were defined in terms of hydroclimate and described by an aridity index (AI)4 provided by the FAO (2006) using climatic variables in the data set CRU CL 2.0		n/a

Table 2.A1.3. Examples of methodologies used to assess water risks (*cont.*)

Water quantity risk indicator	Study	Models used to calculate water demand and supply	Models used to calculate agriculture requirement	Future scenarios considered
Water quality indicators, BOD, N, and P levels	IFPRI and Veolia, (2015)	Using a variety of compiled data sets on population, agriculture, economic growth, climate, municipal wastewater treatment facilities and a newly developed global quality loading model (IGWQLM, IFPRI Global Water Quality Loading Model), the study examines the status of three key water quality parameters – biochemical oxygen demand (BOD), nitrogen, and phosphorus – by estimating their loadings into the water environment in a base period (2000-2005) and in 2050 under six alternative future scenarios focusing on domestic pollution as well as agriculture pollution from livestock and eight key staple crops.		Australia's Natural Science Agency, Commonwealth Scientific and Industrial Research Organisation (CSIRO): medium, optimistic, and pessimistic); MIROC (medium, optimistic, and pessimistic).
Water-related risks on a global scale, focusing upon four headline risks: (1) droughts and water scarcity; (2) floods; (3) inadequate water supply and sanitation; and (4) ecosystem degradation and pollution	Sadoff et al. (2015)	Consequences of hydrological variability for food production were modelled with the IMPACT model, a partial equilibrium agriculture sector model linked with a global hydrology model, a global water supply and demand model, and a gridded global crop simulation model. The model was calibrated to reproduce production averaged over the years 2004-2006.		
Impact of climate change and variability on irrigation Requirements: A Global Perspective	Döll (2002)	The study uses a global irrigation model, with a spatial resolution of 0.5° by 0.5°, to analyse the impact of climate change and climate variability on irrigation water requirements. The study computes how long-term average irrigation requirements might change under the climatic conditions of the 2020s and the 2070s, as provided by two climate models, and relates these changes to the variations in irrigation requirements caused by long-term and inter-annual climate variability in the 20th century.		ECHAM4 climate change scenarios; HadCM3 climate change scenario;

Source: Authors' own work based on the reviewed literature.

Figure 2.A1.1. Water stresses in OECD countries

2013 or latest year available; water abstraction as a % of renewable resource

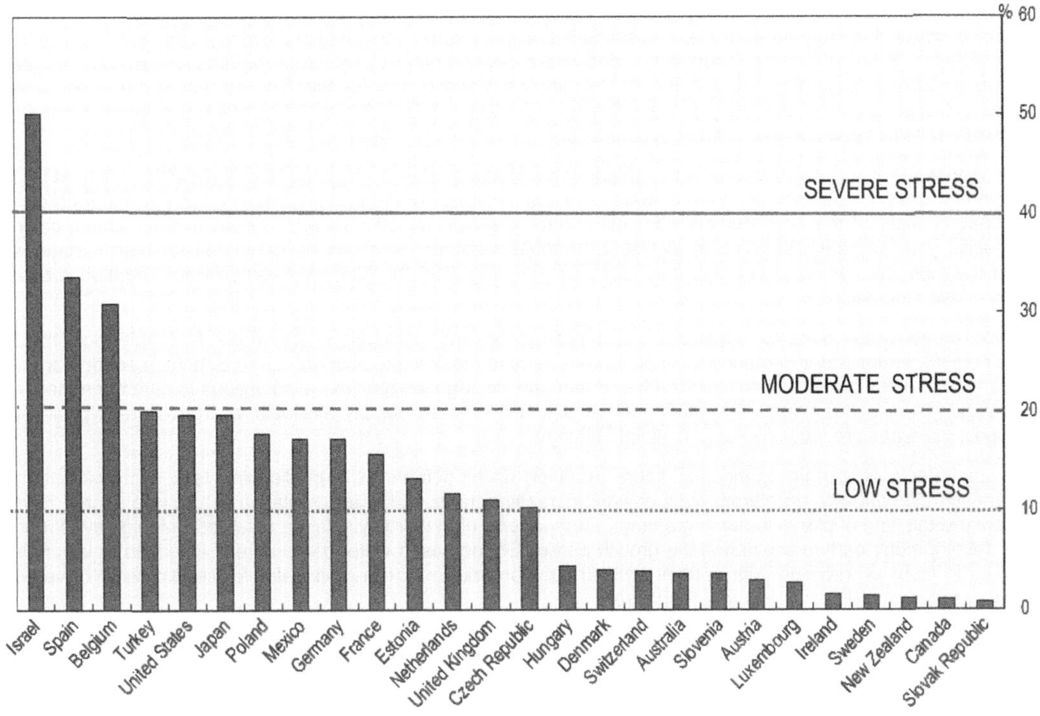

1. Water stress: below 10% = no stress; 10-20% = low stress; 20-40% = medium stress; above 40%: severe stress.
Source: OECD (2015b).

Estimating scarcity water risks for agriculture requires calculating current or future water withdrawal for agriculture. This encompasses assessing current irrigated crops water use and future crop water requirements. Crop calendars, area equipped for irrigation, and water quality also play an important role in the water risk calculation. Box 2.A1.1 summarises the basic steps to estimate water supply and demand.

Box 2.A1.1. Estimating current and future water consumption in agriculture

Water demand calculations rely on estimates of water withdrawal or water consumption. Water withdrawals refer to the volume of freshwater abstraction from surface or groundwater and water consumption to the water withdrawn and not returned to the environment (evaporated or incorporated into a product). Calculations of water quantity take into account water used for agronomic needs and possible changes in water efficiency or water losses in irrigation. It is important to include groundwater abstraction when calculating water quantity risk because groundwater accounts for over 40%of the total consumptive irrigation water use.

Agriculture crop water use is calculated using irrigated crop calendars, actually irrigated crop area, area equipped for irrigation, crop yield data, and irrigation crop consumption of a specific number of crops, or crop group. Irrigated crop calendars provide details on monthly occupation rates of the area equipped for full control irrigation actually irrigated for each crop. These calendars are created for a specific year for which the data is available. Irrigated water demand can be calculated for both surface water and groundwater. For the most part, when calculating water demand from irrigation, the total area equipped for irrigation is used but this can lead to demand overestimation.

To calculate future water risks to agriculture, assumptions on key parameters are made by analysing past statistical relationships. Calculating future water demands for a given area requires taking into account future climate scenarios and estimates on the most likely type of crops grown. Calculating the future demand for irrigation requires assumptions to determine the crops water use efficiency trends. These computations may also need to account for water supply available for agriculture, which depends on competing demands, infrastructure investments and climate change.

Projecting the effects of climate change on future irrigation water demand is a challenging task. Increased temperatures augment evapotranspiration but also accelerate plant growth (increasing the speed of accumulation of "growing degree days", a heat index used to measure the time of plants to reach maturity). Furthermore, higher crop yields are obtained with high evapotranspiration but slower plant development, particularly during the growth period. So increased water demand may also lead to lower yields. The choice of modellers to focus on growing degree days or fixed duration and to look at crop yields for certain water or variable water demand for certain production will lead to diverse responses.

Uncertainties associated with these variables and considerations can result in wide differences in projections. Figure 2.A1.2 shows the projections of 19 simulation exercises on irrigation demand in the future.

Figure 2.A1.2. Irrigations projections from 19 simulations

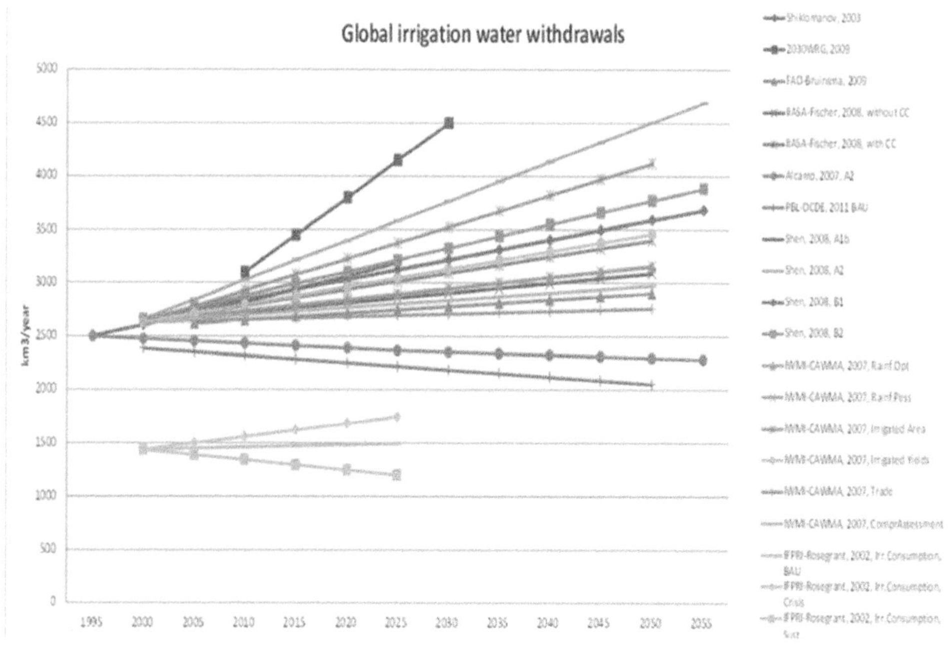

Source: Frenken and Gillet (2012), OECD (2014b).

Multiple indices are used to measure the frequency and severity of extreme water events (OECD, 2016a). Droughts are measured by looking at soil moisture, consecutive dry days or the Palmer index. Flood risks are characterised by the period between two major precipitation events, the total precipitation, and damage costs.

Regarding water quality risks, identifying agriculture production exposed to contaminated water or areas vulnerable to agriculture water pollution is challenging. Globally, the methods used to monitor water quality are poorly developed, especially for pesticides (OECD, 2012). Additionally, differing drinking and environmental water standards between countries hinders global comparative assessment of water quality risks. Groundwater pollution is even harder to document mainly because of the costs involved in groundwater sampling.

2.A1.2. Defining future water risk hotspots for agriculture

Defining future water risk hotspots for agriculture requires the measurement of water risks affecting agriculture, and plausible ways to project these risks in the future. Ideally this could be done by superposing current and future assessments of water risks with expected production areas and activities. Integrated models can help move in this direction (e.g. Schlosser et al., 2014). But the presence of multiple types of water risks, various agriculture activities, and alternative expected futures for both agriculture and water risks may complicate the exercise or increase the uncertainty of outcomes, especially at larger scales.

As a first step, separating the problems into three dimensions may help set up plausible hypotheses that can then be tested with models.[3] Three questions can guide this effort:

- What is the time horizon of the exercise? Current modelling exercises tend to focus either on a medium run (ten years- 2024-2030) or longer term (2050 and/or 2070) when considering climate change effects.
- What is the agriculture outlook looking like in the future under current conditions? The goal is to set a counterfactual of agriculture with reasonable expectations on supply and demand at the decided time horizon. It is important to know production levels by major activities and where these commodities will be produced and distributed. Historic trends can be used to gauge parameters in the future, even if they are generally not sufficient (unless on a short-term horizon) for projections.
- What types of water risks are expected to be observed and where could they coincide with highly productive agriculture areas? Characterising the main water risks may be done qualitatively using past events, trends, climate projections and expert solicitation. On this basis, selected indicators can be used to measure past, current and projected future water risks (see, e.g., section 2.A1.1). Ideally this requires information on potential hazards, exposure and vulnerability. Interactions with agriculture may be multiple, concerning water supply and demand factors, but also the evolution of agriculture management practices.

Figure 2.A1.3 illustrates some of the information necessary to move forward, with an identification of critical future agriculture water risks due to climate change in Europe. It was generated by reviewing the findings of multiple studies on different types of water risks using climate models. Agriculture will face dramatically different risks across the continent. Some regions will face a greater set of risks than others. In particular, the Mediterranean region could be seen as a regional hotspot in view of its multiple challenges: it will face increased droughts, declining water availability, increased irrigation requirements (higher crop water requirements), and declines in water quality and biodiversity impacts. A secondary hotspot may be found in some coastal areas of the Atlantic region that could face sea level rise, flood risks, as well as increased irrigation requirements. To identify hotspots for agriculture production, such assessment would then need to be combined with an evaluation of future agriculture production hotspots in Europe.

Figure 2.A1.3. Regional agriculture water risk assessment under climate change

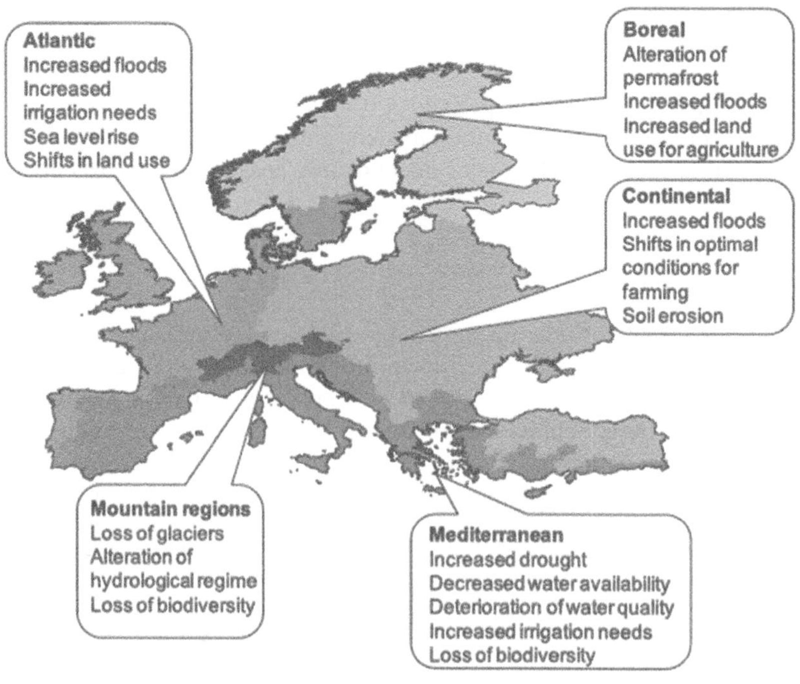

Source: Iglesias and Garrote (2015).

The second step requires setting appropriate thresholds to define the hotspots. Depending on the degree of detail taken in the first-step assessment several options can be considered.

- *Where the future water risks and agriculture projections are well known*, the thresholds can be decided based on the distribution of the risks on agriculture (or the estimation thereof). The threshold cut-off may be defined statistically: above a certain level of risk (or hazard or exposure or vulnerability of agriculture production) that is deemed critical; or relatively by including regions that account for a higher quintile of risks or agriculture production compared to others.
- *Under partial or incomplete information on future water risks hotspots for agriculture*, for instance with an assessment combining multiple studies and datasets that do not all fully agree or that are insufficient to help estimate production levels and risks: the goal will be to look for regions with a consistently higher level of projected agriculture water risks (combined water risks and agriculture importance) compared to others, based on available evidence. The absence of contradiction around the hotspots will also help. Thresholds may be substituted by a selection of few regions in the absence of quantifiable risk estimates. The approach should seek validation, potentially via a consultation with experts from different fields, to avoid costly errors.
- *Where critical information on water risks or agriculture is unavailable, or where information only focuses on a limited area*, a hotspot approach may not be recommended, as it will run the risk of missing hotspots. A general prioritisation exercise may be used instead, considering the entire area (region, country), and allocating investments partially to those areas deemed to be more subject to risk.

Threshold levels are proposed for illustrative purposes (see Table 2.A1.4).

Table 2.A1.4. Examples of possible thresholds to define future water risk hotspots for agriculture

	Based on a comprehensive assessment (via a self-modelling exercise)	Based on partial assessment (using secondary data)
Water shortage	High category (e.g. W/Q>4)or higher quintile of water stress index and/or higher probability of droughts geospatially defined.	Areas consistently found to be at higher risks (high water indices, higher pollution, or higher probability of extreme events) among modelling studies (present and future).
Water excess	Higher probability of floods (e.g. < 1 in 100 years) geospatially defined.	
Water quality issue	Selection based on geospatially defined surface and groundwater quality projections: leading areas in terms of projected deterioration or lowest quality indices.	
Agriculture targets	Hotspot areas selected based on a superposition of the spatial allocation of future water risk with that of expected water-dependent intensive or productive agriculture activities	Predefined risk areas consistently seen as top 1-5 in terms of production or value of selected agriculture products.

Source: Authors' own work.

This menu of options may not be relevant in the presence of a heterogeneous level of information across locations. If a regional assessment of hotspots relies on good information from one part of the region, and lacks information from another, efforts should be taken to ensure that such information heterogeneity does not affect the hotspot determination. For instance, this may encompass a preliminary comparison of available variables everywhere, or increasing data collection and analysis where information is not as immediately available, or allocating a certain level of risks to all areas without sufficient information. If differences are too large across regions, the hotspot approach may not be worth pursuing.

These steps could be applied to any specific scale, since the hotspot approach focuses on relative risks. However, applying the method at a broader or narrower scale may matter in terms of the magnitude of the impacts. Hotspot regions in water abundant countries are less vulnerable to negative impacts on agriculture than in less water-endowed semi-arid countries. The following subsection will discuss multiple examples of partial or more advanced applications at different scales.

2.A1.3. The use of the hotspot approach in OECD countries

Examples from the literature illustrate the range of methods and scales of application of the hotspot approach to determine current or future water risks for agriculture. The emphasis is on diversity; some examples determine actual points on maps while others focus on larger regions and some multiple risks, and others focus on a specific risk. Most of these studies do not fully assess future water risk hotspots for agriculture, but do capture several dimensions of the problem in their varied assessments.

At the international level, recent efforts have focused on using variants of the water stress index to determine water risks. The World Resource Institute's Aqueduct project has applied multiple indicators to spatially define water risks. In particular, Luo et al. (2015) published a 165 country ranking of water stress for 2020, 2030, and 2040, using projections of water stress indices (based on supply and demand projections). Interestingly, they decompose these risks into their contribution to three sectors: domestic, industrial and agriculture sector.

Figure 2.A1.4 shows the results of their agriculture sector estimations for OECD countries rated at high or very high stress as of 2020 (with a water stress index over 3 or 4). According to this metric, only two OECD countries' agriculture sectors would face high water stresses in 2020 (Israel and the United States), but five other countries would join them in this category by 2040 (Spain, Greece, Chile, Mexico and Turkey). As shown in Figure 2.A1.5, agriculture also accounts for the highest component of the overall water risk of OECD average figures. Furthermore, Luo et al. (2015) results (Figures 2.A1.6 and 2.A1.7) suggest that on average OECD countries' agriculture sectors would be less at risk of water shortages under what they define as "pessimistic" climate scenario (high GHG emissions projections) than under an "optimistic" scenario (low

GHG emission projections) Precipitation could increase in many OECD countries in such pessimistic scenario (see, e.g. Ignaciuk and Mason D'Croz, 2014).[4]

Figure 2. A1.4. Estimated water stress indices in the agriculture sector, leading OECD countries, 2020-40

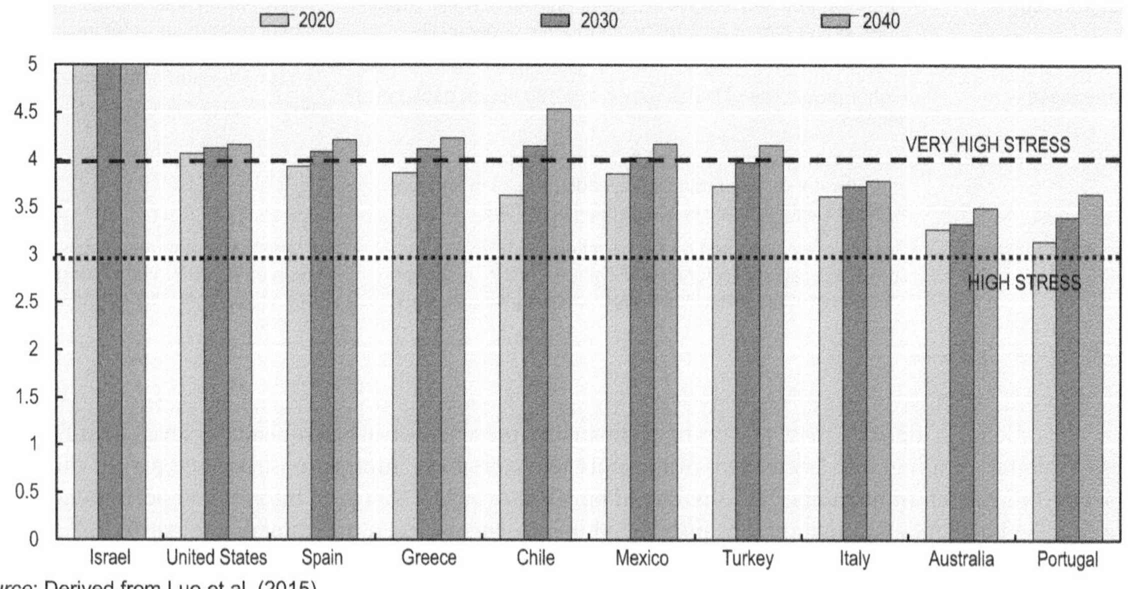

Source: Derived from Luo et al. (2015).

Figure 2.A1.5. Average water stress index for OECD countries under the Luo et al. (2015) pessimistic scenario

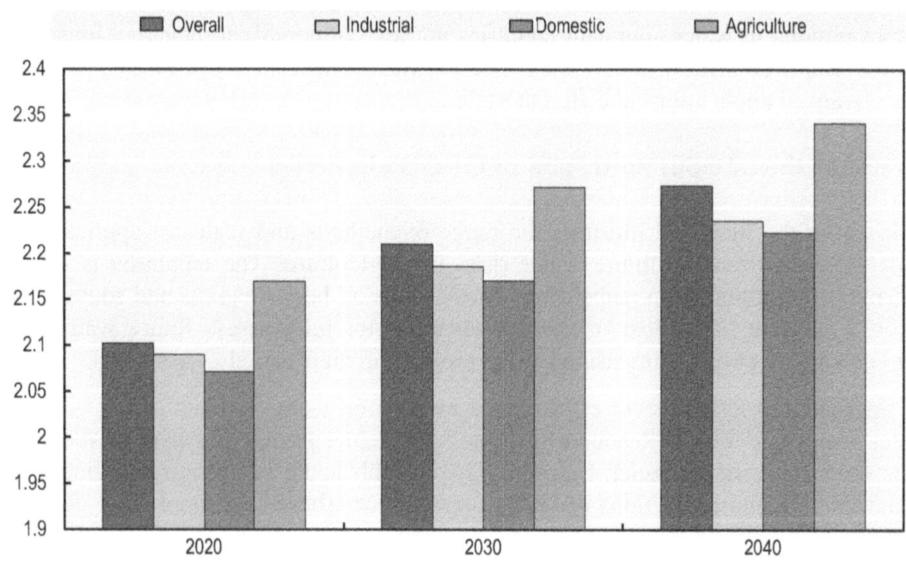

Source: Derived from Luo et al. (2015).

Figure 2.A1.6. Average water stress index for agriculture in OECD countries under Luo et al. (2015) scenarios

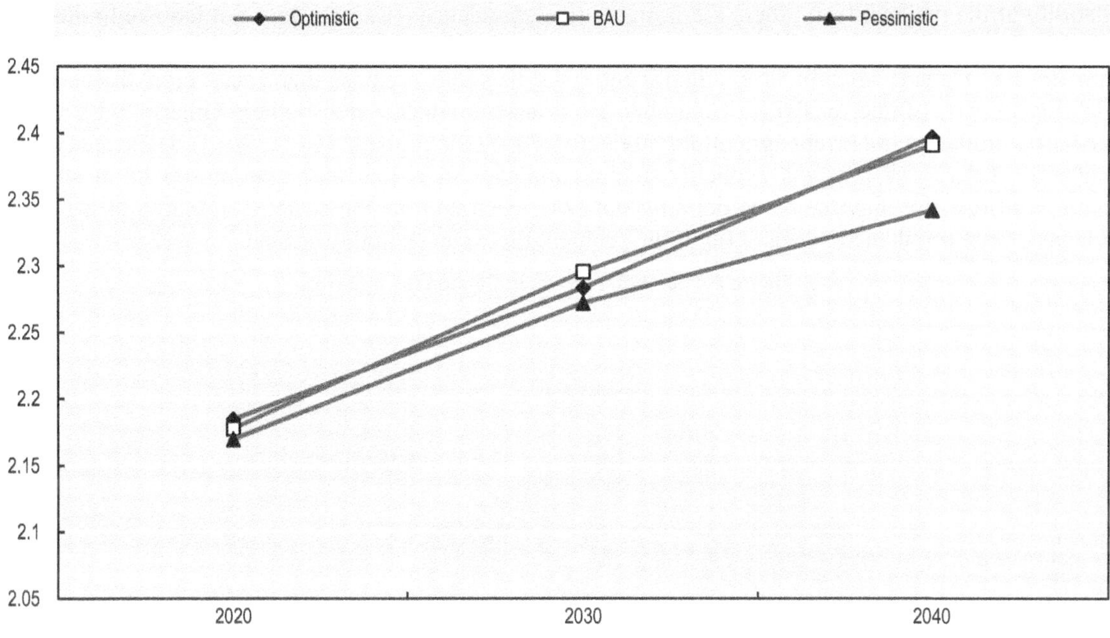

Source: Derived from Luo et al. (2015).

Figure 2.A1.7. Future water stress for agriculture for the top 10 (left) and bottom 10 (right) OECD countries under three scenarios

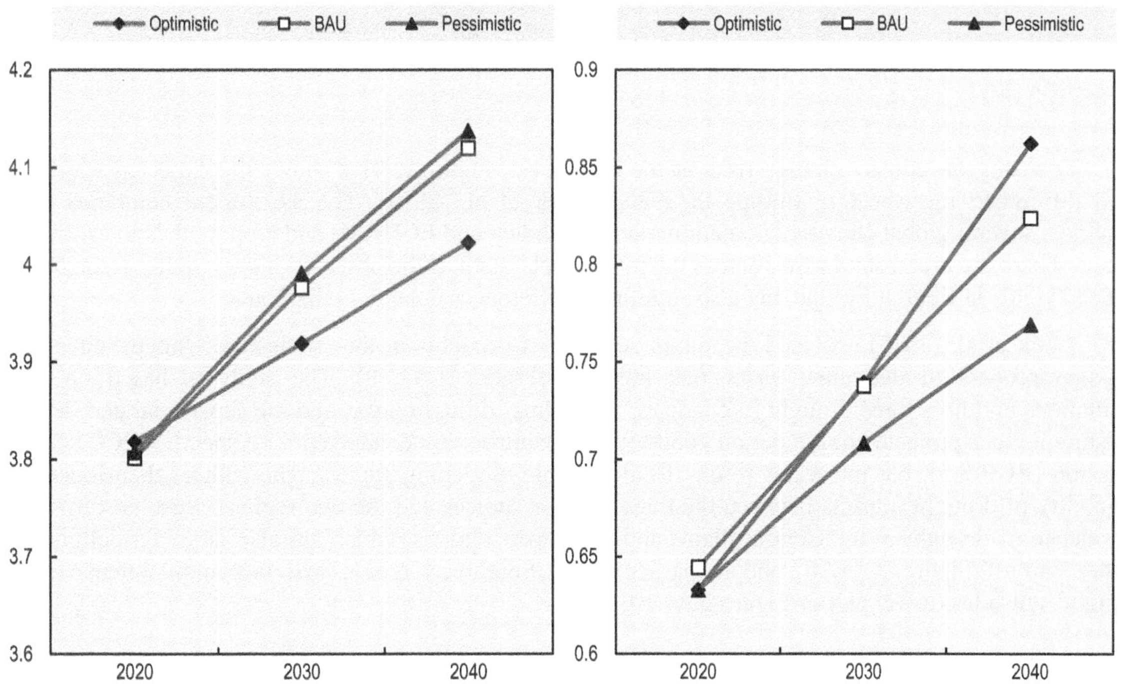

Source: Derived from Luo et al. (2015).

At the national level, multiple countries have conducted water stress assessments, but fewer explicitly consider agriculture. Figure 2.A1.8 shows the example of Canada's water stress. Despite its abundant water availability at the national level, there are limits to this abundance: "Canada does not have unlimited blue or green water available for agriculture expansion or intensification" (CCA, 2013). Furthermore as shown in Figure 2.A1.8, Canada already faces significant risks in some regions; there are "high threats to water availability in parts of interior British Columbia, the Prairie provinces, and southern Ontario, with significant water-based limitation to current agriculture productivity in some regions" (CCA, 2013). If the study does not consider the future risk hotspots explicitly, it does indicate areas that may continue to be at stress. This example confirms the importance of defining hotspots at an appropriate scale. A research at a large scale could mask more problematic areas- even if the problem there can be highly critical.

Figure 2.A1.8. Water stress in Canada (as of 2013)

Source: CCA (2013).

Focusing on climate change risks at the crop level, Figure 2.A1.9 shows the expected water-limited yield difference for wheat in Europe in 2030 (Ciscar et al., 2014). The assessment combines emission scenarios, several global climatic circulation models (Hadley and ECHAM) and crop models to estimate yield effects. There are broad similarities across the two models, such as the observed higher yields in the North and reduced yields in Central Europe, but also significant differences notably in the South.

Cook et al. (2015) analysed the future severity of droughts in the continental United States, using a two-step process to determine water risk hotspots (Figure 2.A1.10). The study is based on multiple simulations and uses three drought risk indices: the Palmer drought index and the superficial and deeper soil moisture indices projected to the period 2050-99. The exercise was conducted for a specific IPCC scenario of emission (RCP 8.5), but integrates results from multiple modelling studies. The authors then looked at the variability of drought impacts between the three drought indices and the two regions. Results show a higher prevalence of drought in the Central Plains and Southwest regions, which are also large agriculture regions (upper part of Figure 2.A1.10) and in particular the Southwest region will face more intensive "mega-drought" episodes (lower part of Figure 2.A1.10).

Figure 2.A1.9. Water-dependent yield effects of climate change on wheat in Europe in 2030

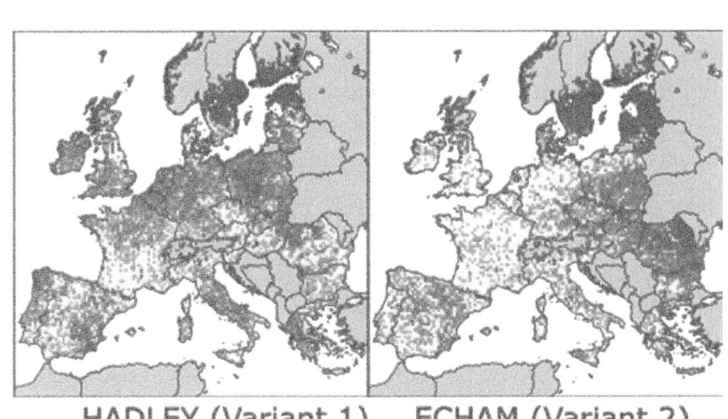

Source: Ciscar et al. (2014).

Figure 2.A1.10. Projected drought indicators in North America, 2050-99

Top panel: Drought maps using the three indicators (Palmer drought severity index, superficial and lower soil moisture) for North America; Middle and lower panel: evolution of these indicators for the Central Plains and Southwest regions of the United States

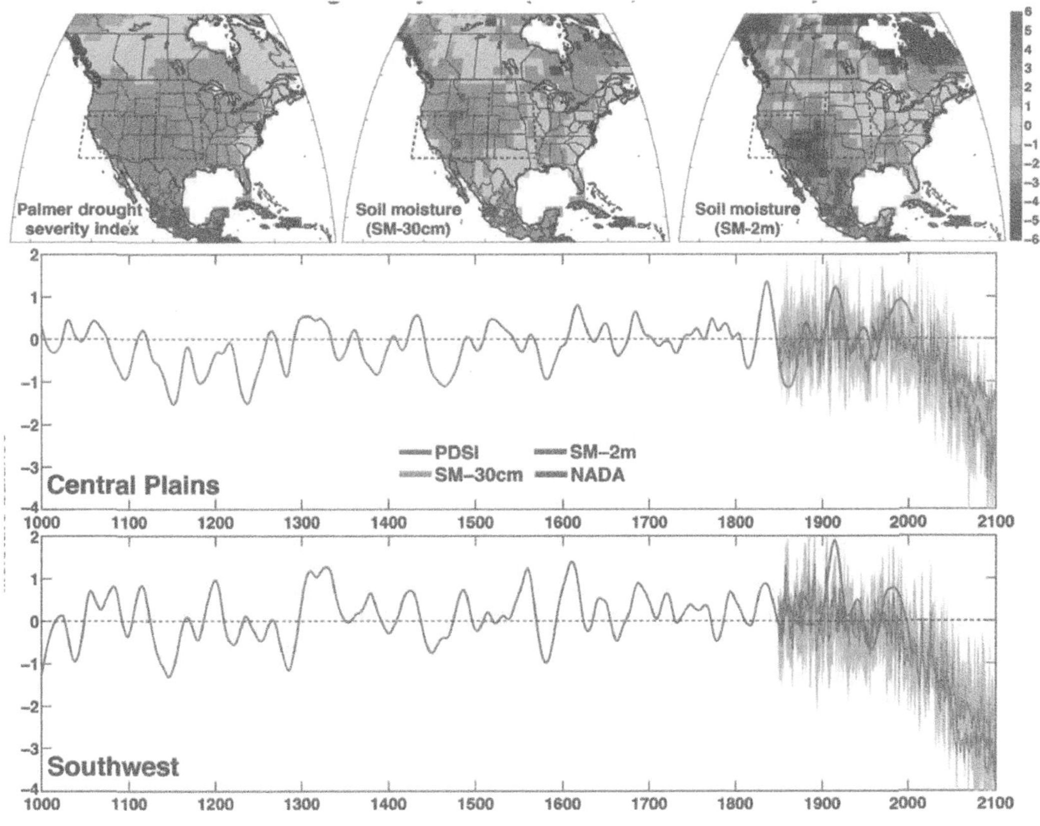

Source: Cook et al. (2015).

In the Flanders region of Belgium, the Government of Flanders has launched programmes to identify areas at risk of floods and droughts. It identifies local hotspots ("signaalgebieden") for flooding, or areas where the geography does not match with the expected areas sensitive to water overflow. The Government of Flanders is also developing a model for droughts, based on a series of indicators, to be supported by a soil moisture monitoring network. Existing data is available, but there is a low awareness and use of data; more generally the risk of droughts should be more widely explained.

Critical water quality issues are also subject to risk mapping and hotspot identification exercises. As part of an assessment of land and water resources, Australia conducted a geospatial assessment of expected areas with high hazards or dryland salinity risk (Figure 2.A1.11 left panel, Thom et al., 2001). Such salinity results from the mismanagement of water resources often due to land clearance or the presence of shallow rooted plants. These plants let the water drain to aquifers, raising water table, thereby elevating salts to the surface where the water then evaporates. The projection accounted for land use change and climate change. A more recent salinity assessment (Figure 2.A1.11 right panel) shows that the areas found to be at risk in 2011 are consistent with this earlier map (Hatton et al., 2011).

Figure 2.A1.11. Risks of dryland salinity in Australia

Left panel: projected salinity risk hotspots, Right Panel: audit of compliance as of 2011

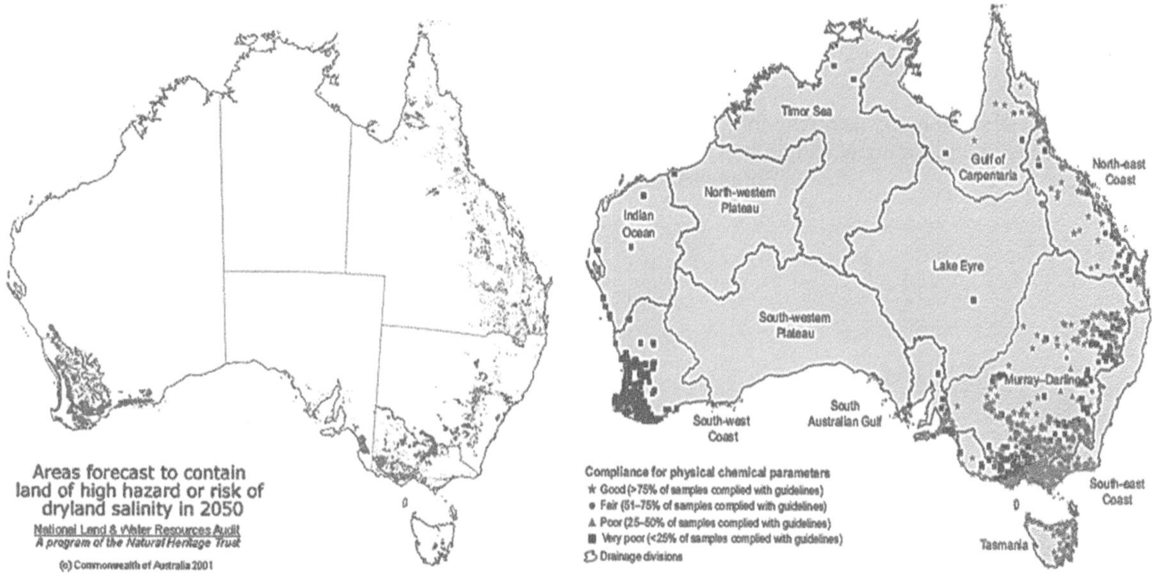

Source: Thom et al. (2001) and Hatton et al, (2011).

The Southland region in New Zealand has used advanced targeting methods to manage water quality damages primarily from agriculture at the regional and local levels. In particular, the Southland region has developed an approach that can help decide which area is the most at risk of water quality impacts (the Mautaura area is shown on the central panel of Figure 2.A1.12), support the characterisation of risks (nitrate infiltration to groundwater, right panel) and customise responses to get the highest results. If this method is used primarily to manage current water risk from agriculture, which may also affect agriculture, it could also be used for assessment of future risks.

Figure 2.A1.12. A physiographic approach to water quality risks in the Southland Region of New Zealand

Left panel: Determination of the physiographic units, middle panel zone at risk in the Mataura PU, right panel mechanism of groundwater nitrate contamination on site

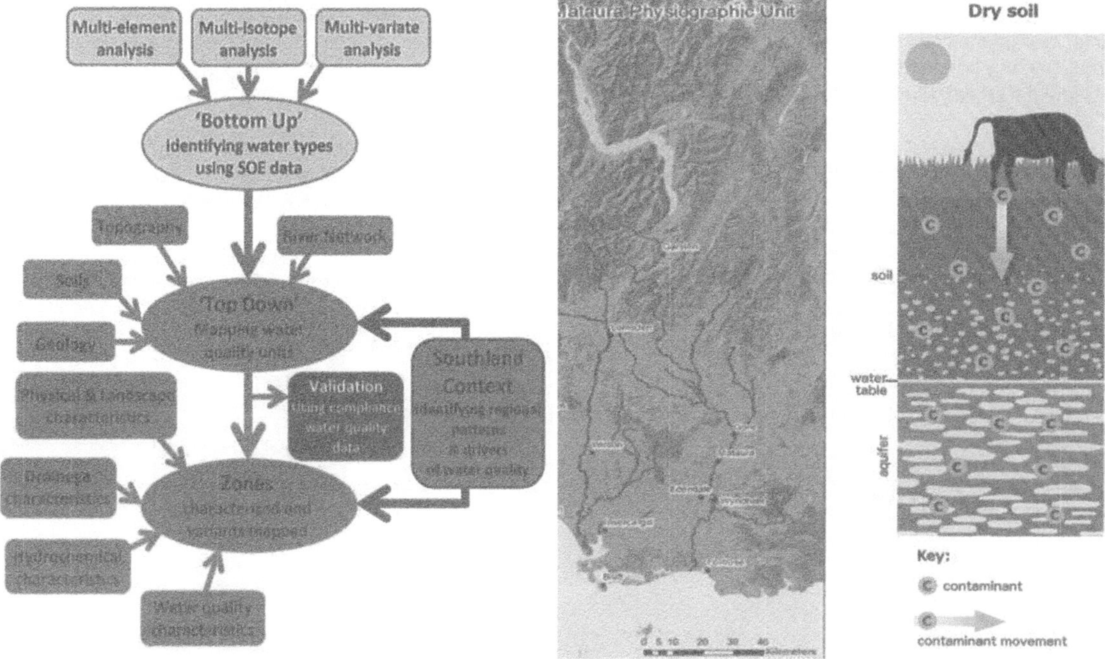

Source: Presentation by B. Chamberlain and P. Ross (2015).

A research project in Switzerland has also used advanced targeting to assess and manage water risks for agriculture in two very specific locations (Fuhrer et al., 2013). As explained in Box 2.A1.2, the project was designed to identify sustainable adaptive management of land and water under climate change. The conclusions highlight the benefits of a highly targeted approach. Although Switzerland is generally considered a water-abundant country, Switzerland is expected to face localised agriculture water risks. The case study hotspot approach is helpful to understand how local adaptation can be effective at any level, but the authors also note that results from the two regions should not be used for generalisation.

These examples demonstrate both the diversity and usefulness of the hotspot approach to assess water risks in agriculture at a various scales. They also pinpoint some of the challenges to determine future water risk hotspots. Most do not incorporate future projections; however, those that do, focus on particular risks (climate change, generally not competing demand factors) or do not focus on agriculture. Furthermore, most approaches involve a combination or cascade of models, each with different assumptions, datasets and scenarios.

Box 2.A1.2. Adaptive management of land and water to mitigate the risks of climate change: A study of two river catchments in Switzerland

Climate change models suggest that there will be a higher crop water requirement and lower water soil availability in summer months in Switzerland. Agriculture will therefore further call upon irrigation, and water course and lake water may face availability constraints. Management solutions are needed to ensure that agriculture can be maintained under increasing water pressure. Broad-based solutions may not be efficient or effective in Switzerland given the country's mountainous topography, which defines the land's hydrology and the pattern of agriculture activities,. Targeted actions may be more likely to lead to optimal results.

To this effect, two agriculture catchment were selected–the Broye and the Greifensee regions–both with similarities and differences. Both regions' agriculture support crop cultivations (including field crops and potatoes) in the lowlands and dairy production in the higher lands, but the Broye region's catchment broadly uses irrigation to support crops and livestock in summer, while the Greifensee region only uses irrigation for vegetables when necessary.

Figure 2. A1.13. Components of the model used for regional optimisation of land and water resources under climate change

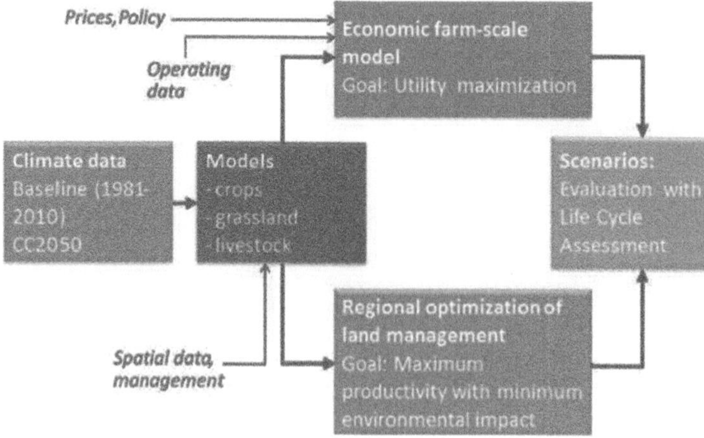

Source: Fuhrer et al. (2013).

A combination of analytical methods was applied to assess current and future constraints and simulate scenarios to respond to the prevalent risks in the two regions (Figure 2.A1.13). A common database with climate information, spatial data, and region-specific management was used to build models at the farm and regional levels that incorporate results from a component model capturing the effect of climate on crop and livestock activities. Life cycle assessment was used to investigate the broader environmental impacts of identified strategies.

The study shows that locally adapted regional plans can help optimise results in the two regions. In the Broye region, this may imply focusing irrigation on the most suitable areas for crops in lowland areas and not irrigating other pasture but instead using crops that are adapted to the climate. The authors also find that at the farm level, there is a trade-off between increased profitability (and often decreasing non-water related environmental impacts) and increasing water stresses regionally. The results further show that (1) changes in farm structures, spatial organisation of crops and farm management practices can, with acceptable production loss, substantially reduce the water needs of a farm, and (2) it is possible to decrease the water needs through implementation of policies such as volumetric water pricing or water quota with relatively small impact on farm income.

Source: Fuhrer et al. (2013).

Annex 2.A2

Additional information on the water risk hotspot selection

2.A2.1. Water risk assessment: Data and method

Measurements of water risks are based on a database of 118 observations from 64 studies. The 64 studies were selected via Google Scholar research (general key words future water risk and agriculture, plus more advanced key words on quality, salinity etc.), snow ball references and additional publications proposed by delegations. The studies are the following : ADRC (2016); Alavian et al., (2009); Alcamo et al. (2007); Arnell (2004); Baettig et al. (2007); Bates et al. (2008); Bijlsma et al. (1996); Brakenridge (2016); Brauman et al. (2016); Cosgrove and Cosgrove (2012); de Sherbinin (2014); Döll (2002); Döll (2009); Elliott et al. (2014); Fischer et al. (2007); Fraser et al. (2013); Frenken and Gillet (2012); Fung et al. (2011); Gassert et al. (2013); Gerten et al. (2011); Giorgi (2006); Gleeson et al. (2012); Haddeland et al. (2014); Hanasaki et al. (2008); Hejazi et al., (2014); Hirabayashi et al. (2013); IFPRI and Veolia (2015); Ignaciuk et al. (2015); IPCC (2012); Jiménez Cisneros et al. (2014); Kiguchi et al. (2015); Liu et al. (2014); Luck et al. (2015); Luo et al. (2015); Mekonnen and Hoekstra (2016); Mendelssohn et al. (2012); Milly et al. (2005); Murray et al., (2012); Nicholls and Cazenave (2010); Nicholls and Tol (2006); Nicholls et al. (1999); OECD (2015c); Parish et al. (2012); Pfister and Bayer (2013); Piontek et al. (2013); Reager et al. (2016); Ringler et al. (2011); Rockström et Karlberg (2009); Rockström et al. (2009); Rowley et al. (2007); Sadoff et al. (2015); Scheffran and Battaglini (2011); Schewe et al. (2013); Shen et al. (2010); Shen et al. (2014); Siebert et al. (2010); Van Drecht et al. (2009); Van Puijenbroek et al. (2014); Villholth et al. (2014); Vörösmarty et al. (2000); Wada et al., (2012 and 2013); WWF (2016); and 2030 WRG (2009).

Observations all report the degree of water risk of different regions or countries at the global scale using different methods. Studies include measurements of changes in: climate change instability hotspots, climate vulnerability index, consumption to Q90 ratio, cumulative absorption/demand ratio, cumulative supply to demand ratios, dry days, emissions of BOD, N and P, extreme storms, flood damages, flood frequency, flood risks, groundwater depletion, human vulnerability to climate change induced decreased groundwater resources; irrigation water demand, irrigation requirements, irrigation water reliability, N and P sewage emissions; coastal land loss from sea level rise, nutrient discharge levels, population, water use and climate impacts, precipitation intensity, pressure on water resources, probability of warm, streamflow, regional climate-change indices, renewable water abundance; runoffs, sea level rise affecting agriculture land, soil moisture, threshold temperatures for significant effects on discharge, water-demand supply gaps, water availability of green plus blue water, water scarcity; water stresses indices (multiple); wet or dry years; withdrawal to available water ratios.

The normalisation of measurements is done by categorical ranking of leading countries at risk to reflect the purpose of the hotspot identification exercise (only areas most at risk matter). The diversity of measurements, therefore, does not prevent the analysis; instead, it allows both covering multiple risk dimensions and increasing the robustness of the identification exercise. The underlying assumption of the assessment is that if a country or region is found to be among the top future water scarcity risks by multiple indices, it is more likely to be facing such water risks, than if a few studies using one method or projection find it at risks.

Care was taken to only take results from scenarios with no simulated response as much as possible (e.g. water risk adaptation or mitigation of risk), so the assessments generally use business as usual or no action scenarios.

Countries and economies with separated measurements are: Afghanistan, Algeria, Angola, Argentina, Armenia, Australia, Austria, Azerbaijan, Bangladesh, Belarus, Belgium, Benin, Bhutan, Plurinational State of Bolivia, Botswana, Brazil, Bulgaria, Burkina Faso, Burundi, Cambodia, Cameroon, Canada,

Central African Republic, Chad, Chile, China (People's Republic of), Colombia, Costa Rica, Côte d'Ivoire, Cuba, Czech Republic, Democratic People's Republic of Korea, Democratic Republic of the Congo, Denmark, Djibouti, Dominican Republic, Ecuador, Egypt, El Salvador, Equatorial Guinea, Eritrea, Estonia, Ethiopia, Finland, Former Yugoslav Republic of Macedonia, France, Gabon, Georgia, Germany, Ghana, Greece, Guatemala, Guyana, Guinea-Bissau, Honduras, Hungary, Iceland, India, Indonesia, Islamic Republic of Iran, Iraq, Ireland, Israel, Italy, Japan, Jordan, Kazakhstan, Kenya, Korea, Kuwait, Kyrgyzstan, Lao People's Democratic Republic, Latvia, Lebanon, Liberia, Libya, Luxembourg, Madagascar, Malawi, Malaysia, Mali, Mauritania, Mexico, Republic of Moldova, Mongolia, Morocco, Mozambique, Myanmar, Namibia, Nepal, Netherlands, New Zealand, Nicaragua, Niger, Nigeria, Norway, Pakistan, Panama, Papua New Guinea, Paraguay, Peru, Philippines, Poland, Portugal, Qatar, Romania, Russian Federation, Rwanda, Saudi Arabia, Senegal, Serbia, Slovak Republic, Slovenia, Somalia, South Africa, Spain, Sri Lanka, Sudan, Sweden, Switzerland, Syria, Chinese Taipei, Tajikistan, Tanzania, Thailand, Togo, Tunisia, Turkey, Turkmenistan, Uganda, Ukraine, United Arabic Emirates, United Kingdom, United States, Uruguay, Uzbekistan, Bolivarian Republic of Venezuela, Viet Nam, Western Sahara, Yemen, Zambia, and Zimbabwe.

For each observation, countries ranked in the high or highest category are considered at risk (in the text and/or in a map or figure), and attributed a value of one for the index representing the specific measurement. This is done for 142 countries found to have some measurement of risks (and ASEAN, OECD, and Mediterranean countries). For observations that target groups of countries, large regions or continents, a value of one is attributed to all the countries in the region.

The individual observations are then summed by category of risk (shortage, quality etc.) and type of measurements (present, past) for each country for the entire database. These numbers are then divided by the relevant total observations to obtain the relevant indicator (total, current, future water risk, etc.). They are then crossed by agriculture shares of commodity production from the IMPACT and AgLink-Cosimo projections.

2.A2.2. Supplementary data from the hotspot identification exercise

Table 2.A2.1. Number of studies reporting specific water risks by country in the reviewed studies

	Total future	Total shortage future	Total excess future	Total variability future	Total quality future
China (People's Republic of)	59	43	13	2	1
United States	53	42	6	3	2
India	49	34	13	0	2
Algeria	42	39	1	2	0
Mexico	40	36	0	3	1
Morocco	39	36	1	2	0
South Africa	39	36	1	1	1
Australia	36	29	5	1	1
Pakistan	36	28	6	0	2
Chile	35	34	1	0	0
Israel	35	34	1	0	0
Tunisia	35	32	1	2	0
Iran	34	33	1	0	0
Jordan	34	33	1	0	0
Lebanon	34	33	1	0	0
Libya	34	31	1	2	0
Spain	34	31	0	3	0
Syria	34	34	0	0	0
Argentina	33	24	8	0	1
Turkey	33	32	1	0	0
Egypt	32	28	2	2	0

Greece	29	28	0	1	0
Iraq	29	28	1	0	0
Kuwait	29	28	1	0	0
Namibia	29	26	1	1	1
Peru	29	21	8	0	0
Afghanistan	28	27	1	0	0
Brazil	27	22	4	0	1
Italy	27	24	1	2	0
France	26	24	0	2	0
Saudi Arabia	26	22	4	0	0
Mozambique	25	20	3	1	1
Qatar	25	21	4	0	0
United Arab Emirates	25	21	4	0	0
Uzbekistan	25	24	1	0	0
Viet Nam	25	13	11	1	0
Uganda	24	18	4	1	1
Canada	23	11	8	3	1
Senegal	23	18	3	1	1
Yemen	23	20	3	0	0
Zimbabwe	23	21	0	1	1
Armenia	22	21	1	0	0
Kazakhstan	22	21	1	0	0
Thailand	22	12	9	1	0
Azerbaijan	21	20	1	0	0
Cambodia	21	9	11	1	0
Ethiopia	21	13	6	1	1
Georgia	21	20	1	0	0
Indonesia	21	8	11	2	0
Russia	21	11	9	1	0
Tajikistan	21	20	1	0	0
Portugal	20	20	0	0	0
Turkmenistan	20	19	1	0	0
Eritrea	19	12	5	1	1
Guatemala	19	15	1	3	0
Honduras	19	15	1	3	0
Kenya	19	12	5	1	1
Kyrgyzstan	19	18	1	0	0
Madagascar	19	14	3	1	1
Nicaragua	19	15	1	3	0
Tanzania	19	12	5	1	1
Venezuela	19	16	2	1	0
Angola	18	13	3	1	1
Costa Rica	18	14	1	3	0
Malawi	18	16	0	1	1
Myanmar	18	6	11	1	0
Nigeria	18	12	4	1	1
Somalia	18	11	5	1	1
Botswana	17	15	0	1	1
Bulgaria	17	17	0	0	0
Burkina	17	12	3	1	1
El Salvador	17	13	1	3	0
Germany	17	14	1	1	1

Former Yugoslav Republic of Macedonia	17	17	0	0	0
Panama	17	13	1	3	0
Philippines	17	7	9	1	0
Colombia	16	9	6	1	0
Rwanda	16	10	4	1	1
Western Sahara	16	12	4	0	0
Benin	15	11	2	1	1
Bolivia	15	12	3	0	0
Burundi	15	9	4	1	1
Djibouti	15	9	4	1	1
Ghana	15	10	3	1	1
Lao PDR	15	7	7	1	0
Malaysia	15	4	10	1	0
Mauritania	15	11	4	0	0
Sudan	15	11	4	0	0
Bangladesh	14	4	9	0	1
Cameroon	14	9	3	1	1
Chad	14	8	4	1	1
Cote d'Ivoire	14	9	3	1	1
Gabon	14	9	3	1	1
Togo	14	10	2	1	1
Zambia	14	12	0	1	1
Central African Republic	13	8	3	1	1
Democratic Republic of the Congo	13	9	2	1	1
Ecuador	13	7	6	0	0
Guinea-Bissau	13	9	2	1	1
Mali	13	9	4	0	0
Niger	13	8	5	0	0
Papua New Guinea	13	4	8	1	0
Romania	13	13	0	0	0
Korea	13	5	5	3	0
Equatorial Guinea	12	7	3	1	1
Poland	12	10	1	0	1
United Kingdom	12	8	3	1	0
Belgium	11	10	0	1	0
Japan	11	3	5	3	0
Netherlands	11	10	0	1	0
Serbia	11	11	0	0	0
Austria	10	8	1	0	1
Czech Republic	10	8	1	0	1
Hungary	10	8	1	0	1
Slovak Republic	10	8	1	0	1
Slovenia	10	8	1	0	1
Switzerland	10	8	1	0	1
Ukraine	10	10	0	0	0
Nepal	9	7	1	0	1
Paraguay	9	8	1	0	0
Sri Lanka	9	4	4	0	1
Uruguay	9	2	7	0	0
Guyana	8	6	1	1	0
Moldova	8	8	0	0	0
Cuba	7	5	0	2	0

Liberia	7	4	3	0	0
Norway	7	1	6	0	0
Denmark	6	3	3	0	0
Sweden	6	1	5	0	0
Chinese Taipei	6	2	3	1	0
Bhutan	5	2	2	0	1
Dominican Republic	5	3	0	2	0
Finland	5	2	3	0	0
Democratic People's Republic of Korea	5	1	2	2	0
Luxembourg	5	4	1	0	0
Belarus	4	4	0	0	0
Estonia	4	1	3	0	0
Latvia	4	1	3	0	0
Ireland	4	4	0	0	0
Mongolia	3	1	2	0	0
New Zealand	2	0	1	0	1
Iceland	0	0	0	0	0

Source: Review of the 64 studies.

Figure 2.A2.1. Future agriculture water risk indices by OECD country, 2024-50 average (8 commodities)

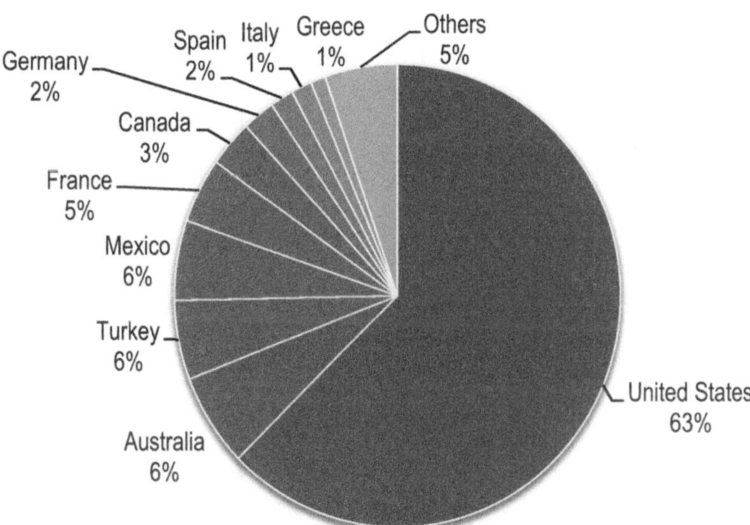

Source: Author's own work based on computations using AgLink-Cosimo, IMPACT and the review of 64 publications.

Figure 2.A2.2. Future water risk hotspot by commodity (2050) (1)

Top left: Beef, Top right: Dairy, Middle left: Coarse grains, Middle right: Rice, Bottom left: Wheat, Bottom right: sugar

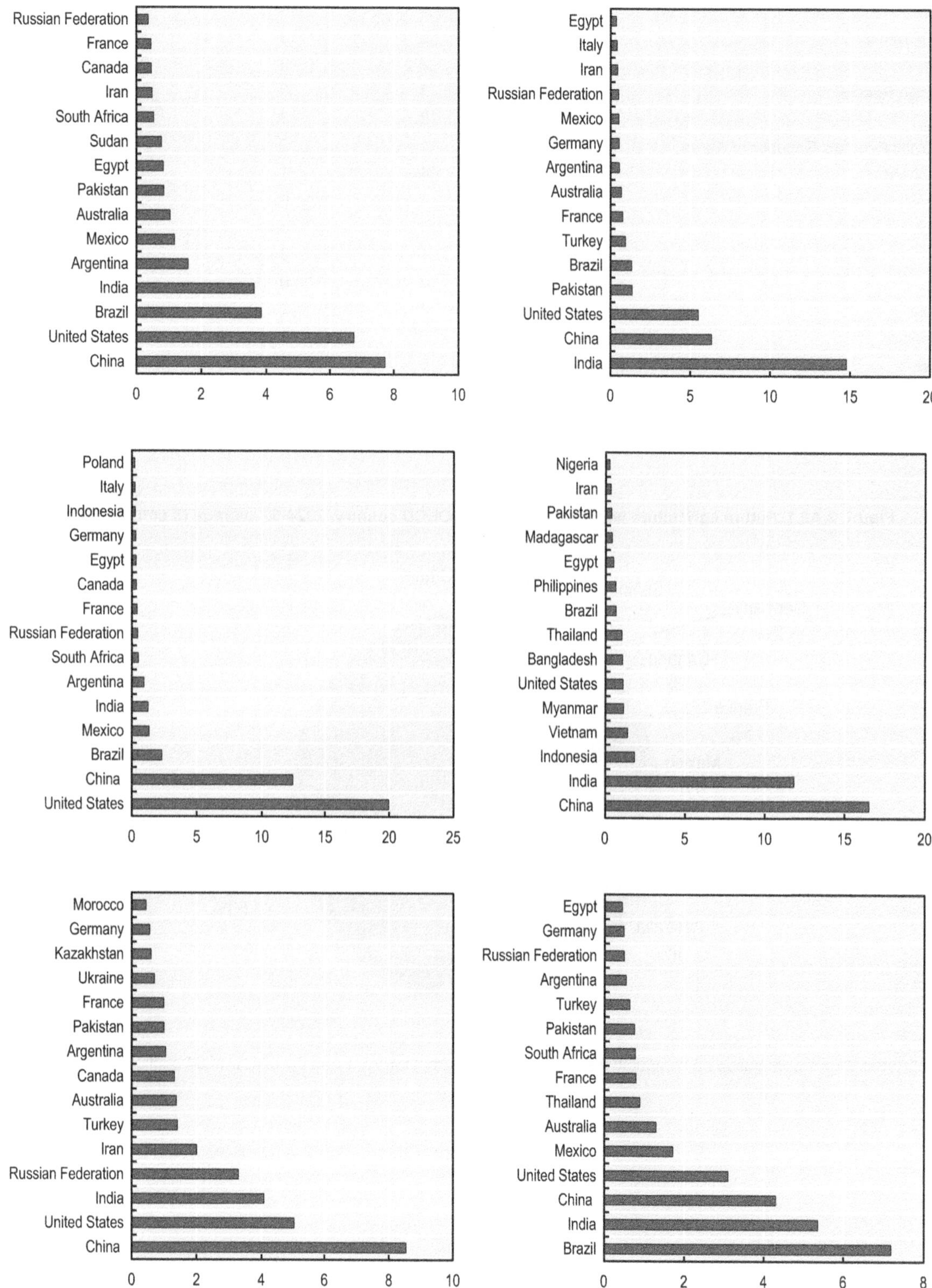

Source: Author's own work, derived from IMPACT projections and the water risk analysis.

Figure 2.A2.3. Future water risk hotspots by commodities (2050) (2)

Top-left: cotton, top right: oilseeds, bottom: fruits

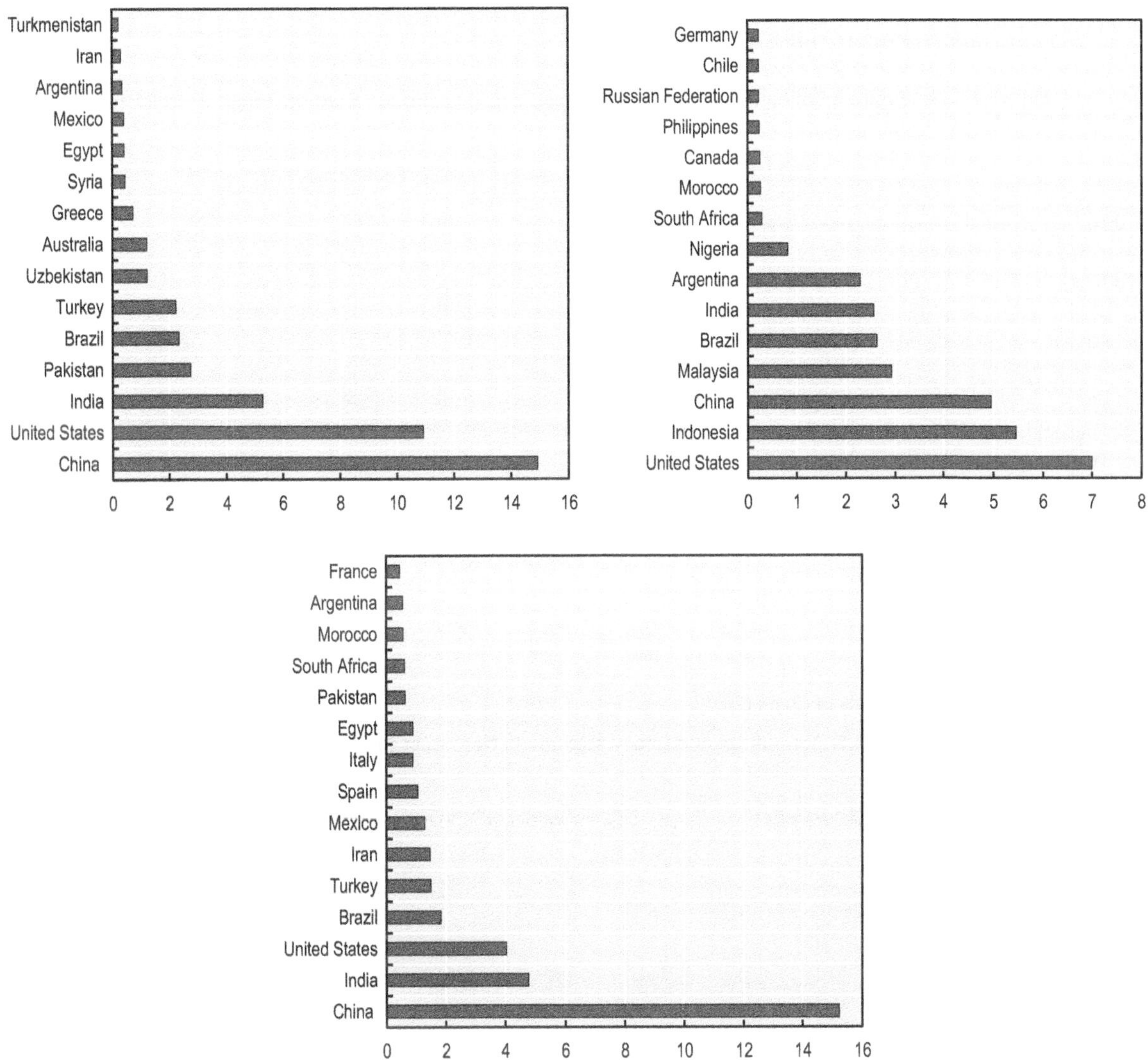

Source: Author's own work, derived from IMPACT projections and the water risk analysis.

2.A2.3. Supplementary information on Northeast China and Northwest India[5]

The case of Northeast China

Recent trends on water and agriculture

Figure 2.A2.4. Evolution of estimated water resources in China and the Northeast Region, 2004-14

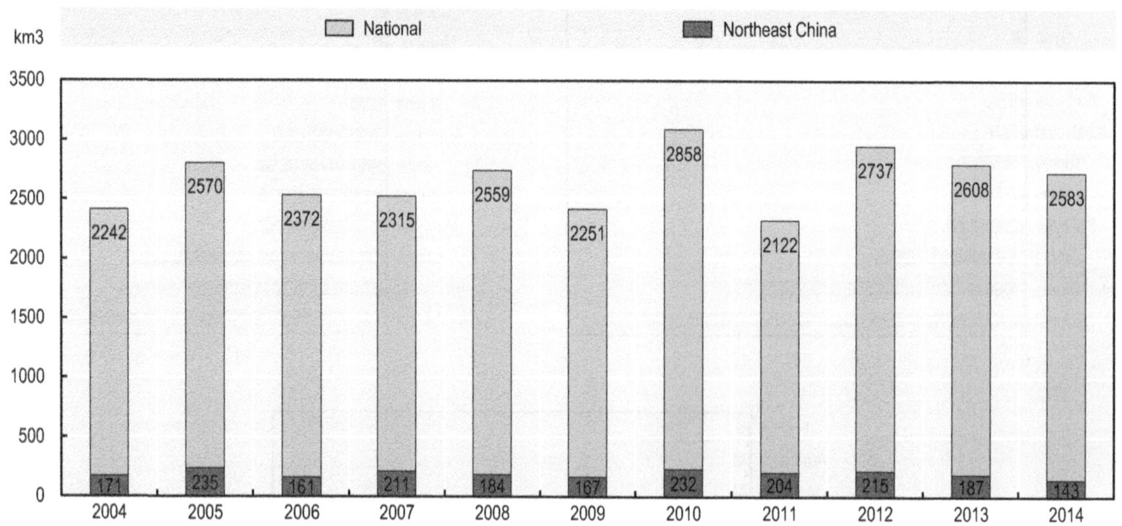

Source: National Bureau of Statistics of China (2016).

Figure 2.A2.5. Agriculture and total water use in Northeast China, 2004-14 (km3)

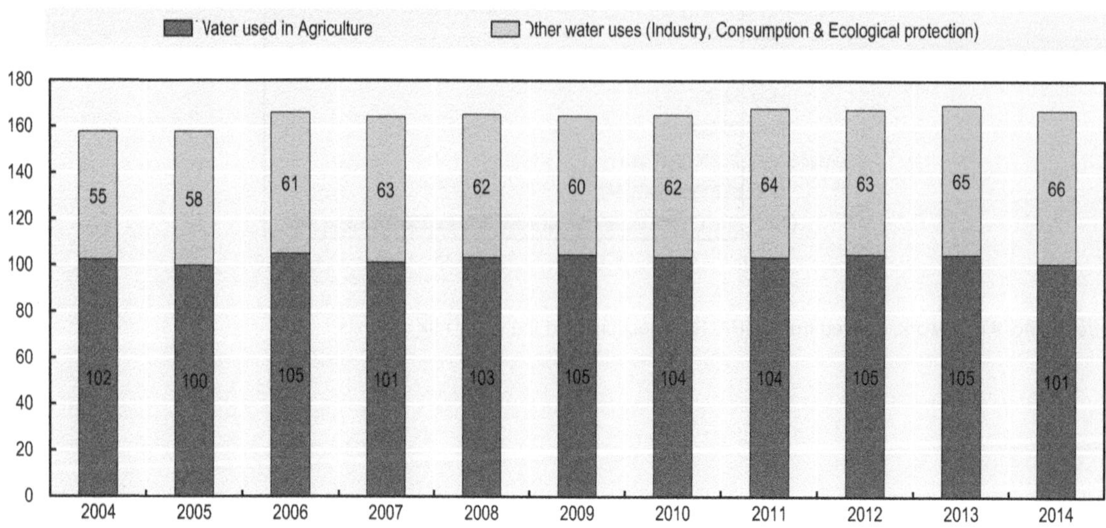

Source: National Bureau of Statistics of China (2016).

Figure 2.A2.6. Yellow River water quality, 1985, 1993 and 2001

Source: Giordano et al. (2004).

Figure 2.A2.7. Regional share of production by crop for Northeast China, 2004-13

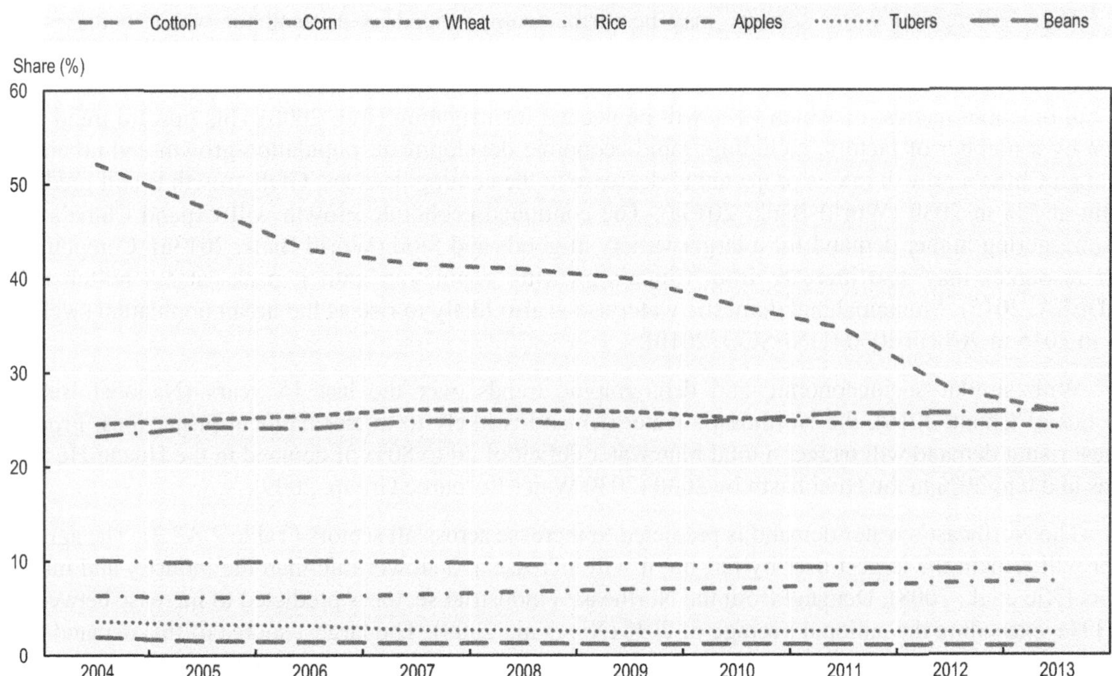

Source: National Bureau of Statistics of China (2016).

Water conditions in the mid to late century

A first determinant of the Northeast's future water resources–climate change–may have varied effects. On the one hand, higher precipitation levels may ease water scarcity concerns over the next century. Some climate change projections point towards increase of summer precipitation (Piao et.al, 2010). In the Huang-Hai Plain (Beijing, Tianjin, Hebei and Shandong) annual precipitation is projected to increase 200 to 300 mm (A2 scenario) or 150 to 240 mm (B2 Scenario) by 2085 (Thomson et al., 2006).

Rising temperatures on the other hand may have a negative effect on water availability due to increased evaporation. Temperatures have increased 0.5°C to 0.8°C over the last century (Ding et al, 2007; PRC, 2007) and are projected to continue to rise 2-4°C in Asia (2046-2065) and 4-6°C in the long-term (2081-2100) (Hijoka et al., 2014). For the Huang-Hai Plain, maximum and minimum temperatures are projected to increase 0.5°C to 2°C by 2030 and 2.5°C and 5°C by 2085 (Thomson et al., 2006). Evaporation for the Huang He basin is predicted to increase between 15 and 49% in 2080 (Zhang et al., 2007).

Water supplies may also be affected by climate change through waning glaciers. 5% to 27% of the area covered by glaciers in China is predicted to disappear by 2050 and 10% to 67% by 2100 (NARCC, 2007). Both the retraction and seasonal melting of these glaciers are expected to impact freshwater levels in Northeast China. In the short run, it will likely contribute to a surge in annual runoff from major rivers. In the long run, water storage may be significantly reduced. Additionally, the rise in sea level could accelerate– increasing water salinity in rivers and aquifers in the (Buckley, 2015). This could imply increases in salinity and land losses.

Estimations of the Northeast's future water supply decrease further if water quality declines. As industry expands, water quality is expected to further deteriorate due to rising pollution levels. China's pollution levels are predicted to increase; high emissions of chemical oxygen demand (COD) are projected to reach 100 million tonnes by 2050, 80 million tons more than the 2030 targets (World Bank, 2013a). As 70% of the rural North China Plain's groundwater is already too polluted to be suitable for human use (China Water Risk, 2016) future pollution is a serious concern.

The challenge of water scarcity may be further compounded as demand for water increases across China; water demand is projected to increase 532 billion cubic metres (61%) from 2005 to 2030 (2030 Water Resource Group, 2009). China's water deficit specifically for agriculture demand in 2030 is predicted to be 43.1 billion cubic metres, of which 91% will be needed for irrigation (Lui, 2006).This upward trend may be driven by a number of factors, including: rapid economic development, population growth and urbanisation. Although China's growth rate is projected to decrease in the coming decades, GDP growth is still predicted to remain at 5% in 2030 (World Bank, 2013a). The continued economic growth will expand China's middle class, triggering higher demand for a larger variety in goods and food (World Bank, 2013a). Competition for water resources may also increase from China's growing population until it peaks at 1.4 billion in 2030 (UNDESA, 2015). Municipal and domestic water use is also likely to rise as the urban population swells from 56% in 2015 to 76% in 2050 (UNESCO, 2010).

With similar socioeconomic and demographic trends over the last 15 years (National Bureau of Statistics of China, 2016), the Northeast's water demand is likely to mirror national projections. Projections suggest rising demand will trigger a total blue water deficit of 20 to 80% of demand in the Huang-He and Hai basins and 0 to 20% in the Huai basin by 2030 (2030 Water Resource Group, 2009).

The Northeast's water demand is predicted to increase across all sectors (Table 2.A2.2). The agriculture sector will remain the largest user by far, but it will increase at a slower rate than the industry and municipal sectors (Xie et al., 2008). Demand from the Northeast's industrial sector is predicted to increase between 24% and 39%, exceeding the national average of 20% (Xie et al., 2008). The largest source of this demand in most of the Northeast is currently thermal power cooling, which may decline given the Chinese government's plan to increase nuclear, wind and solar power generation by 2050. The total water reliance of planned installed capacity is projected to decrease from 100% in 2005 to 70% in 2050 (China Water Risk, 2015). Municipal and domestic demand is also predicted to increase at an accelerated rate. Although the smallest share of water demand comes from the municipal and domestic demand in 2000, it is predicted to almost reach up to more than 20% in some areas by 2030 (Xie et al., 2008). This growing demand for municipal and industrial use will further increase the water stress in the agriculture sector.

Given declining surface water availability and rising demand in the Northeast, current groundwater depletion seems unrecoverable (Cao et al., 2013). Expansion of groundwater irrigation area has continued in the North, with the share of groundwater irrigation rising from 58% in 1995 to 84% in 2012 (Wang, 2017). Intensification of groundwater use has also resulted in lowering water tables. A study of 400 villages in northern China shows for instance that the share of villages with water tables dropping by more than 25cm a year increased from 49% to 63% from the period 1995-2004 to 2006-2016 (Ibid.).

Table 2.A2.2. Current and projected water demand by sector and basin

River basins	Total demand (km3)			Per capita demand (m^3)		Share by sector (%)					
	2000	2030	increase	2000	2030	2000			2030		
						Municipal	Industry	Agriculture	Municipal	Industry	Agriculture
Huai	65.1	71.6	6.5	332	320	10	16	74	18	20	62
Huang	43.7	48.1	4.4	397	364	7	14	79	13	19	68
Hai	40.2	42.9	2.7	312	262	13	17	70	21	21	58
Liao	19.6	22.7	3.1	356	355	13	18	69	20	25	55
China	581.2	653.5	72.3	461	432	10	20	70	16	24	60

Source: Xie et al. (2008).

At the same time, global simulations suggest that groundwater could recharge more than 30% by 2050 in several regions including northern China (Döll and Fiedler, 2008; Taylor et al., 2013). Wide-scale surface irrigation as well as future increases in precipitation could contribute to the recovery (Döll and Fiedler, 2008). Applying improved groundwater management strategies could increase storage by 50m3 by the end of 2030, although 50 more years would be needed for full recovery. If most pumping comes from recharge in the future, groundwater depletion could be avoided (Cao et al., 2013).

Future water availability may have varied implications for future agriculture production

Though limited in scope, a range of models have assessed the impact of certain changes in future water availability on agriculture production. Models focused specifically on climate change suggest that future increases in precipitation may benefit some crops, but the parallel increase in temperatures may outweigh these gains for others. Other models that do not account for climate change, suggest that water stress will increase and negatively impact production across the board. The challenge in aggregating these results is that different assumptions limit comparability across models (Wang et al., 2014). Most studies do not account for factors such as increasing pests and diseases due to higher temperatures, the impact of adaptation measures (Piao et al., 2010), and the possible effect of CO2 fertilisation (Xiong et al., 2010). Most importantly, declining resources due to deteriorating water quality and rising demand are not included in most models (e.g. Chavas, 2009; Chen et al., 2016).

Keeping these limitations in mind, water scarcity is likely to remain an important constraint for agriculture productivity growth. On the supply side, climate change may benefit certain crops through rising precipitation, but hurt others through rising temperatures. The reduction in long term water storage in glaciers from climate change may also negatively impact future agriculture production in the long run. Moreover, future increases in run-off may exacerbate soil erosion and thus reduce water retention; the loss of top soil, which contains about 75% of vital plant nutrients (Young, 1989), is a key concern for agriculture production. Furthermore, declining water quality in the Northeast may further constrain water access for agriculture production. Of course, some relief may be provided by infrastructure projects such as the South-North water diversion project to increase water supplies in the North.

On the demand side, rising competition for water resources is likely to negatively impact agriculture production. Urbanisation and a growing population with higher incomes will not only increase the direct demand for water but also for water-intensive industrial goods and food products. Rising incomes may also change food consumption patterns towards highly processed goods, such as meat, and thus may motivate farmers to increase water-intensive production of livestock. On the other hand, water efficiency gains—due to improvements in on-farm irrigation technology as well as efficiencies of scale through China's shift towards land consolidation—may ease production requirements for the agriculture sector.

Though not all of these factors are taken into account—studies generally do not account for water demand from other sectors, groundwater use, or other issues—several projections are also available for specific crops, with mixed results (Wang et al., 2014). Chavas et al. (2009) suggest that rice, wheat and maize yields will increase significantly; however, this result could be inflated by their assumptions about CO2 fertilisation effects (Hare et al., 2011). Other studies paint a more nuanced picture. According to Hijoka et al. (2014), rising precipitation will not be enough to offset the negative impact of higher temperatures on rice yields. Tao et al. (2009) predict that corn yields in the North China Plain will also decrease (by 9.1% to 9.7% by 2020, 15.7% to 19.0% by 2050 and 24.7% to 25.5% by 2080). However, wheat production, especially winter wheat, could increase in the Huang-Hai Plain due to rising temperatures and precipitation (Hijoka et al., 2014; Xiong et al., 2010). Without these positive climate effects, however, wheat could be negatively affected in the future (OECD, 2015c). Given the high productivity rates and low water requirements, future potato production may also increase (Chavas et al., 2009).

The case of Northwest India

Agriculture and groundwater depletion

Punjab and Haryana are agriculture strongholds in India. The two states account for only 3% of the national territory, but produce 15% of its rice and 30% of its wheat (Ministry of Statistics and Program Implementation, 2016c). They achieve the highest yields in the whole country both for rice and wheat. A hectare of cultivated land in Haryana or in Punjab produces more than twice more wheat as an average hectare of cultivated land in India (Figure 2.A2.8). In the two states, the income of the majority of the population relies directly on agriculture (Bhupal, 2012; Punjab Directorate of Agriculture, 2016).

Groundwater use is intensifying in Punjab. Until the end of the 1990s, the share of croplands under tubewell irrigation in Punjab increased rapidly. Simultaneously, yields for major crops entered a period of stagnation (World Bank, 2013b; Rang et al., 2014). At the beginning of the 2000s, the area under groundwater irrigation hit a plateau, and even slightly decreased between 2001 and 2012 (Figure 2.A2.9). In spite of this, groundwater development[6] continued to rise. Thus, the recent worsening of the groundwater depletion is not generated by an extended access to tubewell irrigation, but by the growing intensity of water pumping (Figure 2.A2.10).

Figure 2.A2.8. A productive grain production region

Top: Average yields (kg/ha) for major crops in Punjab and Haryana, compared to nationwide; Bottom: evolution of average rice yields in Punjab

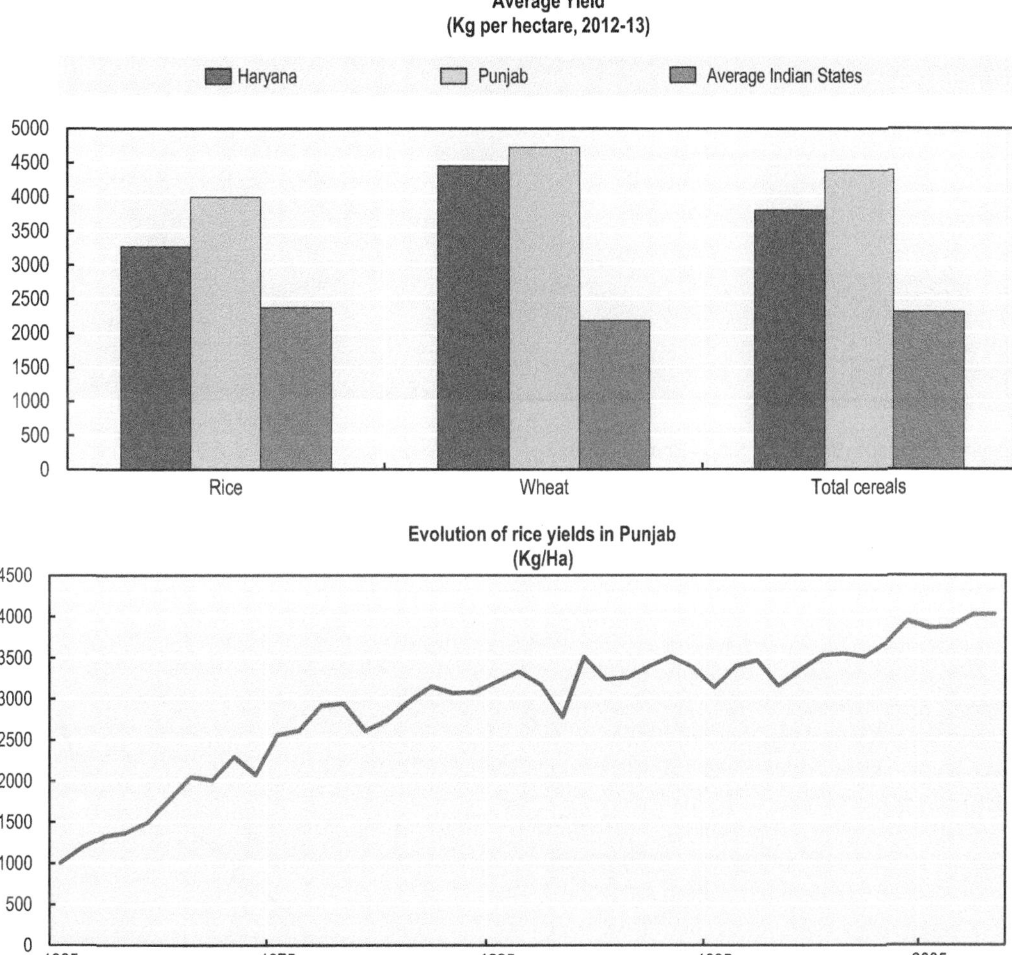

Source: Ministry of Statistics and Program Implementation (2016c); Rang et al. (2014).

Figure 2.A2.9. Evolution of irrigated area by source in Punjab, India

Source: Agricultural Census; Singh and Singh-Bhangoo (2013).

Figure 2.A2.10. Proportion of area with groundwater depth exceeding 15m in Punjab, India

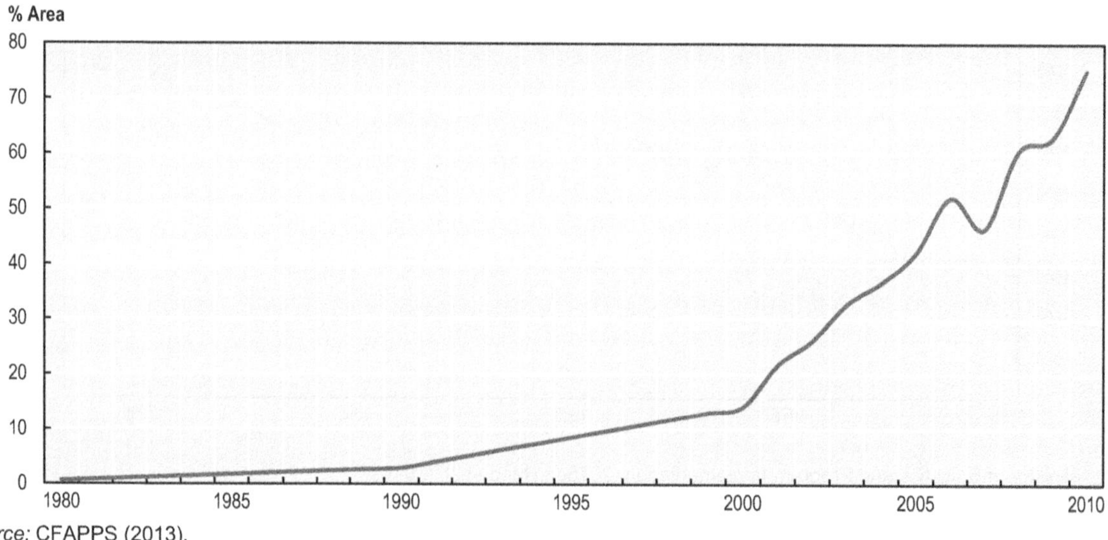

Source: CFAPPS (2013).

Expected water situation in the next decades: A growing water supply-demand gap

While most of the region is already under water stress and groundwater resources are overexploited, water demand is still expected to grow in the upcoming decades. If the cropped area seems to have reached a plateau in this region, and a further expansion of net irrigated area is unlikely (GIST Advisory, 2013), demography dynamics, urban and industrial development will put further pressure on water allocation. In India, total water demand is expected to increase by 32% by 2050 (Amarasinghe et al., 2008). Changes in temperature and precipitation patterns could also increase the groundwater needs for irrigation.

By 2050, the industrial and domestic sectors will account for 84% of the additional water demand over India (Amarasinghe et al., 2008). Punjab is currently the State with the lowest share of non-agriculture groundwater use. Haryana is the third lowest. However, with a rising population and changing consumption patterns, domestic and industrial uses are taking greater importance. In the next ten years, the region will gain

5.8 million new inhabitants, and the urban population will rise by 43% (National Commission on Population, 2006). Between 2004 and 2011, non-agriculture draft already rose from 2% to 6% of total groundwater use in Haryana (CGWB, 2006), and by 2025 Punjab projects a rise by 38% for domestic and industrial groundwater consumption compared to 2011 (CGWB, 2014).

Climate change projections point towards extreme heat and water scarcity, with land suffering higher soil moisture deficits and increased evapotranspiration, which will increase the need for irrigation during the summer and the Rabi seasons (World Bank, 2013b; Bruinsma, 2003). Temperatures are projected to rise from 1.5 °C to 2.5°C by 2030, and by 2 °C to 6.5°C by 2080 in Haryana according to various scenarios (CCAFS and CIMMYT, 2015). Global and regional climate projections show more intense precipitations in Northwest India, which will increases the risk of floods (Döll, 2002; Gosain et al., 2001), but the adverse effects caused by higher temperatures and changes in rain concentration patterns will overwhelm the potential benefits of increased precipitation over the year (Döll, 2002; IPCC, 2008).

Due to its semi-arid climate, the Indo-Gangetic plain is naturally exposed to salinity risks. Intensive groundwater pumping extents and amplifies the phenomena, which limits irrigation sources for farmers in affected areas (CGWB, 2015). According to BGS (2015), increasing salinity caused by groundwater abstraction and intensive irrigation in the region could be a bigger threat than falling water tables. Salinity limits water uptake capacity of plants, and dramatically decreases yields for most of the crops (Shrivastava and Kumar, 2014). Almost 0.5 million hectares are already affected by salinity in the State of Haryana, and this surface is expected to grow in the coming decades, following the patterns of groundwater depletion (Kim, 2013). In Punjab, salinity problems from the Southwest (on a surface representing 40% of the State) are now extending to the central districts (Mahajan et al., 2012). By 2023, in some central districts of Punjab, water tables will sink beyond 50m depth. Today the groundwater of these regions is still fresh, but forecasts estimate that water salinity will increase because of pumping-induced reverse flows coming from surrounding brackish aquifers (Hill-Clarvis et al., 2016; Mahajan et al., 2012). In addition, the mobilisation of deeper and more saline water through tubewell irrigation affects the quality of shallow waters (BGS, 2015). Finally, there is evidence of pollutant breakthrough and water leakage to the deep reservoirs of the multi-layered aquifers, due to intensive pumping (Lapworth et al., 2014).

Future industrialisation also raises concerns about development of groundwater pollution. Indeed, sewage water is commonly used for field irrigation and fertilisation; but in Punjab, 60% to 70% of industrial effluents are discharged in sewage drains without any treatment. Therefore, in the vicinity of large cities and factories, soils and groundwater get contaminated with anthropogenic pollutants such as mercury, lead, and other toxic metals (Aulakh et al., 2009). Domestic and drinking purposes are the first hit by these contaminations and in some areas, groundwater has also been declared unfit for irrigation (Singh, 2001). Groundwater monitoring authorities support recycling of industrial effluents as a way to minimise both groundwater withdrawal and pollution (Pandey, 2011).

Climate change projections announce significant changes that will adversely affect the hydrological situation in southern Asia. In the future, intra-seasonal climate variability will be exacerbated, and the risk of unexpected drought will increase (World Bank, 2013b). Empirically, the latest unusually dry cropping seasons have led to higher groundwater uptakes to compensate surface water and soil moisture deficit. Therefore, aquifers will be more and more solicited as reliable water reservoirs to answer hydric stress (Krishan et al., 2015; Mohinder, 2016; IPCC, 2008). In addition, increased temperatures will decrease the water storage efficiency of surface water reservoirs and open irrigation canals, that is to say the capacity of these systems to conserve and deliver water without loss (CGWB, 2013). Besides, the Indus basin and the Ganges basin are broadly supplied by snowmelt water (PSCST, 2014). Since melting mountain glaciers are declining on the long term, rivers flows will be affected during the summer as early as 2050 (Barnett et al., 2005; World Bank, 2013b; PSCST, 2014). Finally, even if precipitations will increase in total, they will be more variable and concentrated in time during a shorter monsoon. As a result, groundwater recharge could drop if storage capacities are not improved (Bruinsma, 2003; Taylor et al., 2013).

Notes

1. Consequences may also be positive, but the word is more widely associated with a negative outcome.

2. IPCC (2012) defines resilience as "the ability of a system and its component parts to anticipate, absorb, accommodate, or recover from the effects of a hazardous event in a timely and efficient manner, including through ensuring the preservation, restoration, or improvement of its essential basic structures and functions".

3 The order of the assessment is inter-changeable; the determination of water risks may precede agriculture area, so long as they are coinciding at the end.

4. This, however, could increase flood risks.

5. Supplementary information on the case of the Southwest United States can be found in Cooley et al. (2016).

6. Groundwater development is defined as the current annual rate of groundwater abstraction divided by the mean annual natural groundwater recharge.

Chapter 3

Water risk hotspots and the impact on production, markets and food security

This chapter analyses the impact of future water risk hotspots on agriculture. It reviews agriculture production impacts, considers the market and trade effects of such risks, and looks at the broader effects of these risks on food security. The discussion combines insights from the literature on water risks in agriculture, from case studies on Northeast China, Northwest India, and the Southwest United States, as well as results from a simulation of the global impact of projected agriculture water risks in these three regions.

Key messages

Agriculture water risks at hotspot locations can cause three layers of impacts. First, water risks will directly affect agriculture production. Second, these production effects may have broader market implications both domestically and potentially internationally. Third, broader food security and associated indirect effects may also occur.

Agriculture production effects at hotspot locations will vary in scope and duration according to the type and combination of water risks. In the absence of policy action, Northeast China's agriculture is expected to face water shortage due to competition for water use with other sectors, possible continued groundwater depletion in parts of the region, and the uncertain effects of climate change, leading to possible pressure on field crops such as maize and wheat. In Northwest India, groundwater depletion and quality impacts will combine with more variable monsoons to exert pressure on wheat and rice. In the Southwest United States, continued surface water volatility, higher temperature, increased water demand and groundwater pressure would reduce irrigation acreage and may result in a shift of activities away from feed and forage crops towards activities with higher revenue per amount of water used.

Agricultural impacts in highly productive regions can spillover to domestic and international markets. A simulation of simultaneous gradual irrigation stresses in the three hotspot regions, results in price rises of 5% to 8% of selected commodities globally, reductions in production in the three countries, and significant shifts in trade of field crops, such as maize, wheat, rice and cotton. A drought in the US Southwest alone in 2021 would not significantly affect global markets, but a combined drought in Northwest India and Northeast China in 2030 would result in price increases of similar amplitude and shifts in trade. Simulated global climate change projections alter these impacts, in some cases aggravating those for key crops in the three countries.

Recent developments show that agriculture water risks can also have impacts beyond the sector, affecting food security and indirectly resulting in impacts to other countries. Agriculture water risks partially explain the multiplication of foreign land purchases often by water scarce countries in relatively better water endowed countries. Agriculture water risks can also result in social tensions, fuelling conflicts that can become regional, and they can drive migration both domestically and internationally.

The previous chapters identified specific countries and regions as future water risk hotspots for global agriculture production. This chapter analyses more specifically the impacts such risks may have on agriculture locally, nationally and globally on market partners, and broader food security. The chapter addresses impacts at any water risk hotspot location at the regional, national or subnational levels, even if much of the analysis focuses on the three identified regions of Northeast China, the Southwest United States and Northwest India.

3.1. Ripples and risks: Three layers of impacts from agriculture water risk hotspots

Water risk hotspots for agriculture concentrate risks, but can generate different layers of impact. In relatively well-integrated food markets, high agriculture production risks may impact food chains locally, regionally and internationally. Agriculture production risks may also have different types of economic effects on producers, market chain actors, consumers of the concerned products, and even other consumers if they impact the price of other commodities. These impacts will often be less extreme the further from the maximum risk location (hotspot) and the longer after a shock (if there is a shock), but different types of impacts may also disturb this ripple-like effect.

To capture these gradual and differentiated effects synthetically, three levels of impacts arising from agriculture water risk hotspots (Figure 3.1) are evaluated. A first tier of impacts, mainly agriculture production-related, encompasses direct effects in the targeted (hotspot) regions. A second tier includes market-related indirect impacts, affecting market actors, possibly trade partners, and competitors.[1] The third tier comprises food security and broader consequences, which may concern a larger range of countries. Any country may be affected by one or more of the three types of impacts, or not be affected at all by specific agriculture water risk hotspots.

Figure 3.1. Three layers of impacts from water risk hotspots

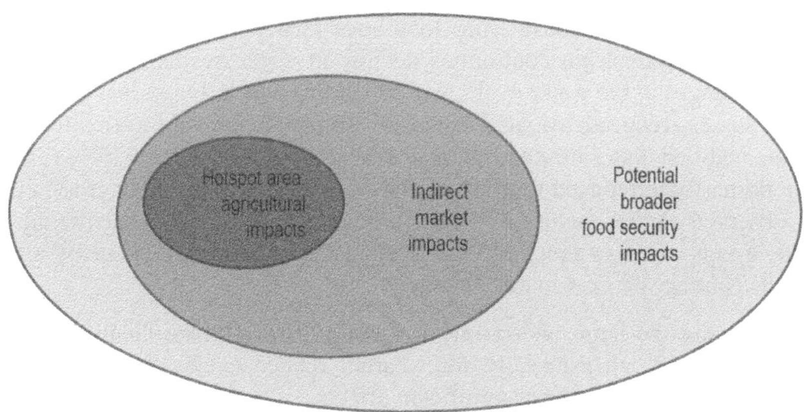

Source: Author's own work.

The following three subsections will use this impact differentiation as a benchmark. For each of the three levels of impact, some of the key expected effects of agriculture water risks hotspots will be reviewed, drawing in particular from analyses touching on the three case studies.

3.2. Local water risks at the hotspot locations: Productivity losses and changes in agriculture activities

Main types of agriculture impacts from water risks

The nature and amplitude of production impacts depend largely on agriculture water risks. Table 3.1 gives an illustration of the type of agriculture impacts one might envisage for specific regions under different time and risk horizons. Expected impacts range from losses of productivity and increased costs of production to observed shifts of production. The type of impact will vary for different water risks; in Brazil water shortages and excess affect depress agriculture income, but shortages have a much stronger effect than floods (Hidalgo et al., 2010). The impacts will also vary by agriculture activity due to the differences in water demand elasticities in the short and longer run, the vulnerability of farm systems, market integration and substitution. Livestock operations will not face the same risks as perennials or annual crops. Implications on land allocation and use will therefore be variable.

Table 3.1. The impact of water risks on agriculture

Type of water risk	Short-term temporary shocks	Medium-term prolonged	Long-term irreversible impacts
Quantity: "too little"	Droughts affecting rainfed agriculture in particular, and putting pressure on surface water irrigation.	Prolonged dry conditions with diminution of surface water and groundwater recharge leading to rainfed agriculture loss, limits on irrigated agriculture, affecting grass-fed livestock affected and putting pressure on groundwater systems.	Loss of rainfed and irrigated land, eliminating extensive livestock systems with limited seasonal surface water and possible severe groundwater depletion and quality issues.
Quantity: "too much"	Severe floods affecting seasonal crops	Progressive augmentation of surface water requiring drainage systems and changes in crop and livestock agriculture activities.	Excess water preventing agriculture activities, surface and groundwater flooding.
Quality: "too polluted"	Eutrophication Salinisation Other contaminations	Increased concentration and costs of treatment, new risks with uncertain treatments.	Unusable water systems-irreversible groundwater or soil contamination.

Source: Author's own work.

Impacts may also vary when considering long-term changes versus increased variability of water availability. For instance, a gradual but continuous decline in water availability, due to the reallocation of water demand or slow changes in the water cycle, will result in a similarly gradual reduction of soil moisture, leading to crop yield losses, reduced irrigated land and ultimately prevent agriculture crop cultivation. In Australia, it was estimated that water-limited potential wheat yields reduced by 27% (or 1.1%/year) between 1990 and 2015, due primarily to reduced rainfalls in the same period (Hochman et al., 2017). This evolution contributed to the observed stalled yields of wheat nationally during the same period. In contrast, a few repeated dry extreme events can also shock the soil system but be followed either by a (partial) recovery of soil moisture, or on the contrary by irreversible changes.

These differences and the large uncertainty associated with climate change make future agriculture impact assessments highly difficult especially for weather related events. Still, reviewing ex post studies provides evidence that these events can have significant effects.

Damages are especially important in developing countries. FAO (2015) evaluated the damages and losses[2] from 140 weather-related disasters affecting more than 250 000 people between 2003 and 2013 in developing countries. It found not only important damages to agriculture (84% of the economic impact of drought), but also strong declines in agriculture outputs. Moreover, economic losses to farmers tended to be transferred on to further losses in the food value chains and the rest of the economy in countries highly dependent on agriculture. Table 3.2 provides an overview of results from FAO studies on the damages and losses of selected weather extreme events to the agriculture sector of mainly developing economies. Of the 78 disasters with available sub-sector damage data, the crop subsector absorbed over 42% of total damages and losses caused by disasters. Livestock activities were slightly less affected, accounting for 34% of the total economic impact of weather-related disasters to agriculture.[3] Furthermore, the losses to agriculture vary by type of disaster. Floods and storms explain up to 93% of the damages from extreme weather events to the crop subsector. Droughts tend to impact more on livestock: 86% of the economic losses from weather-related disasters to livestock are caused by droughts.

Table 3.2. Agriculture damages and losses due to storms, floods and droughts

Type of event	Country affected	Estimated damages	Estimates losses
All extreme weather-related disasters (2003-13)	Developing economies		Losses worth USD 80 billion in crops and livestock production (333 million tonnes of cereals, pulses, meat, milk, etc.) (worth USD 80 billion) Average 2.6 percentage-point loss of agricultural value-added growth
All extreme droughts (1991-2013)	Sub-Saharan Africa		Losses worth USD 31 billion lost in crop and livestock production Food imports increased by USD 6 billion Average 3.5 percentage-point loss of agricultural value-added growth
El Niño southern oscillation event (2015-2016)	Philippines	144 083 ha of farms affected	Losses worth USD 70.8 million in crops (218 379 tonnes of crops lost (worth USD 70.8 million)
***El Niño* southern oscillation event (2015-2016)**	El Salvador	60% of maize crop destroyed	85 858 ha of maize lost or damaged
Typhoon Haiyan (8 November 2013)	Central Provinces, Philippines	USD 700 million damages to the agriculture sector 600 000 ha of farmland (crops, orchards, plantations) affected 44 million coconut trees damaged or destroyed	1.1 million tonnes of crops lost
Typhoon Koppu (18 October 2015)	Central Provinces, Philippines	USD 180 million in agriculture damage	
Heavy rainfall (October 2015)	Ono Island, Fiji		Drop in sugar cane production by 25%
Flood (2010)	Pakistan	USD 5 billion estimated in damages and losses	Agriculture sector growth dropped by 3.3 percentage points between 2009 and 2010 National GDP dropped by 1.2 percentage points
Drought and extreme cold winter (2014-2015)	Mongolia	Area available for grazing pasture reduced to cover only 60% of national herd's needs	Wheat harvest down by 40%
Drought (2008-2011)	Kenya	USD 11 billion estimated in damages and losses	Agriculture sector growth dropped by 5 percentage points in 2008 and by a further 2.3 percentage points in 2009

Source: FAO (2015).

Weather-related events can also generate important damages in developed economies (OECD, 2016). Box 3.1 reports the results of empirical studies on drought agriculture damages. They show that droughts can result in significant crop yield reductions in Europe and North America, but that the specific amplitude of damages is highly dependent on regional differences. Similar estimations did not find significant average effects of floods on crop yields in these regions. This may be partially explained by the fact that agriculture damages from floods are generally not well reported (OECD, 2016).

Box 3.1. Estimating the effects of droughts on cropland

Lesk et al. (2016) estimated the impact of droughts and floods on crop production, area harvested and yields. Droughts tend to reduce national cereal production significantly by 10.1% on average, equivalent to roughly six years of production growth. In particular, drought was linked to significant decreases in both yield and area harvested. However, no significant lasting effect was noted in the years after the disasters as production resumed normal levels immediately after the drought.

When disaggregating the data by region, Lesk et al. (2016) found that droughts had a greater impact on cereal yields, and consequently on total production, in Australia and New Zealand as well as developed economies in Europe and North America, than it did in Asia, Africa and Latin America. Indeed, cereals production was reduced by 19 percentage points in the more developed economies compared with 12.1 and 9.2 percentage points in Asia and Africa, respectively. The data did not show any significant impact of droughts on cereal production in Latin America and the Caribbean. The authors explain this difference by the fact that agriculture in developed economies is more technically developed, with seed varieties selected to make the best of normal optimal weather conditions. When the weather varies from this optimal, the yield potential is not reached. The relatively lesser yield gap from drought in developing economies is due to the larger prevalence of traditional indigenous seed varieties that still have some tolerance to naturally more frequent drought occurrences.

Even within Europe, the agriculture impact of droughts varies highly between countries. Naumann et al. (2015) estimated the relationship between drought and crop production in European countries that have not changed their border in the past 50 years. Using meteorological data to calculate several drought indices and FAO data to track grains production between 1950 and 2012, the authors posit that the relationship between drought intensity s and the average damage to crop production D can be modelled as: $D(s) = \alpha.s^{\beta}$. Declining this formula to different countries, the results show that the expected damage to crop production in Spain and Italy follows the severity of droughts according to a power law. In Germany and Portugal, the relationship between drought severity and damage to crop production is linear. In France, the United Kingdom, Ireland, Denmark, Austria, Hungary and Romania, damages to crop increase at a decelerating pace as drought severity increases. Interestingly, historical data show that crop production in Finland and Norway actually increases with drought severity.

Source: Lesk et al. (2016) and Naumann et al. (2016).

A key question relates to the long-term effects of water risks on agriculture systems: When do water risks become irreversible (see Table 3.1)? The dynamic of water risks determines the potential irreversibility of their impacts and the possible responses from the sector. One could anticipate that agriculture continues to recover under gradually worsening water conditions up to a certain threshold. This could happen, for example, with the continuing depletion of aquifers worsening to the point where it is no longer economically viable to use them for irrigation. However, it is difficult to anticipate such a threshold when crises are observed at different time intervals. For instance, three significant climatic events (droughts or floods) may happen over a few years and the next events may not happen for ten years or more, making it difficult to evaluate whether conditions have irreversibly changed.

Even if their effects may not be irreversible in the long run, extreme weather events that follow each other frequently can have lasting high impacts on agriculture production, threatening local and national food security (FAO, 2015). For example, the agriculture sector of the Philippines was affected by 75 consecutive disasters between 2006 and 2013 for a cumulated damage of USD 3.8 billion; Pakistan's agriculture sector was also struck by recurrent disasters between 2005 and 2011 for cumulated damages of USD 8 billion; and Tabasco State of Mexico witnessed five consecutive floods in the years up to 2011 with damages reaching USD 1.2 billion. When droughts extend over several years, they impose a longer level of stress on agriculture production. Several studies have identified a significant negative impact of long drought episodes on the overall growth of the agriculture and food sector in the countries affected (FAO, 2015; UNISDR and CRED, 2015).[4]

The overall vulnerability of production systems will determine whether repeated crises lead to long-term effects. Because of their overall greater vulnerability to disasters, farmers in developing economies are more challenged than those in developed economies to find the seeds, fertilisers, machinery and technical assistance needed to start a new production season and recover from weather events.

Expected impacts for agriculture production in the three hotspot regions

Water scarcity is likely to remain an important constraint for agriculture production in Northeast China. First, rising competition for water resources is likely to negatively impact agriculture production. Urbanisation and a growing population with higher incomes will not only increase the direct demand for water but also for water-intensive industrial goods and food products. Rising incomes are expected to change food consumption patterns towards meat and highly processed goods. This may motivate farmers to increase production of livestock, which is usually more water-intensive per unit of output than crops and lead to the expansion of local agro-industries, which will contribute to raising demand for clean water. Though it is difficult to draw general conclusions from projections,[5] literature suggests that climate change may benefit certain crops in Northeast China through rising precipitation but hurt others through rising temperatures. On the other hand, higher temperatures may increase irrigation water demand (Wada et al. 2013)[6] and the projected reduction in long-term water storage in glaciers will negatively impact future agriculture production in the long run.

Several studies suggest that the effects associated with increasing future water demand will dominate the effects of climate change on the Chinese agriculture sector., including in the Northeast Combining water demand and diverse climate projections, Rosenzweig et al. (2004) find that "Northeastern China suffers from the greatest lack of water availability for agriculture" among 11 major international agriculture regions. Hezaji et al. (2014), also combining socio-economic scenarios with climate change, show that the increase in water stress in the region is very large in the future, regardless of the implementation of any greenhouse gas mitigation scenario. Tao et al. (2003), modelling both factors, find that the soil-moisture deficit would increase generally with consequent impacts on yields. Xiong et al. (2010) model climate change, water and agriculture land and find that economic drivers supplant climate effects, concluding that "there will be insufficient water for agriculture in China in the coming decades, due primarily to increases in water demand for non-agriculture uses". In part of Northeast China, rice, wheat and maize areas could reduce by up to 60% by 2040 according to their projections. Combining crop and water simulation models with climate change to capture change in climate, water availability and land constraints, Xiong et al. (2009) project a drop of cereal production by 9% (scenario B2) to 18% (Scenario A2) in 2040. None of these projections account for water quality degradation and the continued depletion of the North China Plain aquifer.

In the North China Plain sub-region, the status and use of groundwater reserves will be determinant for the future of cereal production. Irrigated activities will be threatened if the water table continues to fall (as projected in the literature). Grogan et al. (2015) estimates that without the mined groundwater[7] agriculture irrigation currently uses, crop production in the region would fall 101 million tonnes, or 10% of the national production. Additional recharge could increase groundwater if precipitation effectively increases in the future, but the lag for water to reach the aquifer may mean that it comes after much of the stock has been depleted.

Losses in productivity may also have broader consequences on rural development. Future projections suggest that income levels for rice and corn producers are particularly vulnerable. As rural incomes decline, rural to urban migration may also increase as farmers seek alternative income opportunities. The lack of sufficient drinking water and the impacts of extreme events may be additional drivers of migration.[8] By 2020, it is estimated that 30 million environmental refugees will flee water stress in China (World Bank, in China Water Risk, 2016).

The combination of climate change and groundwater depletion is also expected to affect the viability of agriculture in Northwest India. First, climate change projections consider that India's agriculture will be significantly impacted (e.g. OECD, 2015). Due to higher temperatures, rainfall variability and decreasing access to freshwater for irrigation, productivity of most of the crops in India is projected to decline by 10% to 40% by the end of the century (Shrivastava, 2016). In the Northwest region, rising mean temperature and higher frequency of extremely hot days are expected to reduce yield in future decades (Jalota et al., 2014). Considering both factors of moisture deficit and heat, CRIDA (2013) projected that irrigated rice yields would fall by about 16% in Haryana and Punjab by 2050. Wheat could be even more affected, since it is grown during the Rabi (winter-spring) cropping season, which is more exposed to drought, and because it is a winter crop particularly sensible to heat stresses (Krishan et al., 2015). Livestock will also be concerned: the Government of Haryana (2011) forecasts a 10% to 25% loss in milk production due to rising temperature and moisture deficit. Finally, the increase of mean temperatures and the intensification of monsoon rains will favour the development of pests and parasites (Hundal and Kaur, 2006).

At the same time, continued groundwater depletion may decrease farmers' resilience to droughts. Thanks to tube well irrigation, crops in Northwest India are less vulnerable to drought than other regions, ensuring them the lowest inter-annual production variability (Kawashima, 2012). However, the actual pace of groundwater depletion threatens the sustainability of this system. Seckler et al. (1999) estimate that if pumping costs become out of reach for farmers using tube well irrigation, India could lose 25% or more of its total crop production. If Punjab had no access to irrigation, the maximum attainable yield for wheat would fall by half compared to the current local yield, and by more than two thirds for rice (Bruinsma, 2003). A collapse of the groundwater irrigation system would be all the more dramatic in Northwest India since rainfed crops are particularly exposed to climate change in this region. Soora et al. (2013) found that the Northwest will suffer from the highest yield deviation in the country for rainfed rice by 2080, with a decline ranging from 7% to 22% according to various scenarios.

Possible increases in the cost to pump deeper groundwater may exacerbate farmer's losses resulting from lower average yields and production levels. In the coming decades, ICAR (2015) forecasts a 1.78% annual growth rate of energy used by irrigation pumps in the country. Although electricity subsidies reduce the cost impact on farmers—while inducing further depletion (Box 2.4 in Chapter 2)—they will face increased capital expenditures. Low-cost wells equipped with surface-mounted centrifugal pumps are not powerful enough to lift water beyond eight meters (Sekhri, 2014). Therefore, farmers have to invest in deep tube wells equipped with electric submersible pumps, costing around USD 2 500. A survey conducted in Punjab reveals that from 2006 to 2014, 54% of farmers had purchased submersible pumps and 45% invested in higher capacity motors (Walia and Sharma, 2015). From 1997 to 2007, total annual expenses on tube well infrastructures by farmers of Punjab reached about INR 15 billion (e.g. USD 362 million in 2007). At the same time, the aggregated debt of farms rose four-fold (Hira, 2009). Indebtedness and water access restrictions caused by depleting groundwater have dramatic consequences for rural development: in average, villages where aquifers have fallen below eight meters suffer from a 10% increase in poverty rate (Sekhri, 2014).[9]

Changes in water availability in the Southwest United States will affect agriculture because water is an important input for crop production and agriculture is the largest water user in the region. There is a large and growing body of literature on the impact of water scarcity on Southwestern agriculture (Frisvold and Konyar, 2012; Frisvold et al., 2013). These studies show that urban encroachment and the growing demand for water to accommodate continued population growth and restore degraded ecosystems are projected to reduce water availability for agriculture.

The combination of increased water demand, higher temperature and more volatile precipitations is expected to reduce irrigated land and agriculture water use in the region. For instance, the total amount of land irrigated at least in part with Colorado River water could decline from current levels by between 4% and 16% by 2060, primarily due to reductions in irrigated area and land conversion in Arizona (USBR, 2012). Similar reductions are expected in the case of California, where urban growth on agriculture land, with lower water requirements per area, could offset the increased water demand due to increased evaporation (CDWR, 2013 and 2014). Agriculture in the southern portion of the California Central Valley, in particular, could experience the greatest shortages, with reliability (or water supply being able to meet demand) consistently below 95% and below 50% in the hottest and driest climate scenarios (CDWR, 2013). These projections also do not account for expected reductions with the implementation of the 2014 groundwater legislation, which shall reduce groundwater use in the short run, but increase reliability of groundwater resources once implemented.

These effects will have significant impacts on agriculture production. Table 3.3 shows the results of varied projections. Overall, crop production in the Southwest United States will cover less land due to limits on water supplies and urban encroachment. Lower value, water-intensive crops, such as feed and forage crops, are likely to experience the greatest losses. Feed prices are likely to rise, thereby increasing costs for livestock and dairy producers. Additionally, climate change is likely to alter the location and productivity of pasture and rangeland, the distribution of livestock parasites and pathogens, and the thermal environment of animals.

Table 3.3. Examples of projected impacts from climate change and future water stresses in the Southwest United States

Agriculture activity	Region or states	Expected impacts	Reference
All	Southwest	Effects of 25% water reduction: Irrigation area reduced by 1.5%, increased in dryland production area. Land fallowing reduces net income by USD 65m.	Frisvold and Konyar (2012)
All	California	Irrigated land and water use decline by 20% and 21%, respectively, with revenue decline by 11% due to partial offset from higher crop prices and crop shifting. The largest reductions in crop area were for pasture, field crops, grains, and rice.	Howitt et al. (2009)
Field crops	Entire country with specific regions	By 2050, climate change (1) reduces crop yield for all field crops, except wheat and hay; (2) reduces irrigated crop area due, in part, to reductions in surface water availability in the US West; and (3) reduces the production of all crops. Higher commodity prices are insufficient to offset declining crop yields and returns, thereby reducing producer welfare.	Marshall et al. (2015)
Specialty crops	California	Changes in winter chill by 2050 "will no longer support some of the main tree crops currently grown in California", but it will have little effect on almonds and pomegranates because these crops have low chilling requirements.	Luedeling et al. (2009)
		No specialty crop really gains from warming, cherries are at the greatest risk; almonds may suffer from winter warming.	Lobell and Field (2011)
Livestock	Southwest	Warmer temperatures will likely lengthen the growing season and improve pasture productivity at higher elevations and in the North of the region, offsetting increased feeding cost rises.	Frisvold et al. (2013)
		In the South, such as Arizona, and Central and Southeastern California, higher temperatures will shorten the growing season and reduce yields.	Brown et al. (2015)
Dairy	Entire country with specific regions	Diminished dairy productivity due to heat stress: Maricopa County, Arizona, would experience a diminished dairy productivity of 18% by mid-century, compared to a loss of 4% in Tulare County, California, and 2% in Weld County, Colorado.	Mauger et al. (2015)

Source: Authors' own work based on cited studies.

The impacts of these changes on farmer livelihood, rural communities will depend on the magnitude and rate of change, as well as the ability to adapt to these changes. In an economic assessment, Frisvold and Konyar (2012) estimated that a 25% reduction in water availability in the Southwest United States would reduce regional employment by only 3%, with the largest impacts in Arizona. Howitt et al. (2014) projected that the direct, indirect, and induced job losses from the California drought would total 17 100 seasonal and part-time jobs. Actual employment data, however, found that agriculture employment reached a record-high 417 000 people in 2014 (California Employment Development Department, 2015). While agriculture employment would likely have been even higher if there had been less land fallowing in 2014, these losses were offset by a shift away from field crops that employ relatively few people per hectare of land, toward tree crops and tomatoes that employ more people per hectare (Cooley et al., 2015).

3.3. The market and trade effects of water risks in the agro-food sector

From production risks to market and trade impacts

Market responses to agriculture production shocks depend on multiple market variables. Prices do not always respond to changes in yields. They respond when production shocks are large enough. For instance, the yields of corn and soybeans in the Midwest have been found to be negatively correlated with US market prices (Dismukes and Coble, 2006), because, in this case, "most farm-level yields are closely related to area-wide production" and the area's production accounts for a significant share of world production (Ibid.). An important production shock in a non-integrated small market may lead to large price changes in that market. In contrast, large yield shocks in a limited area of a relatively open country or in a country with a small global market share will not impact prices.

Market impacts also depend on supply and demand. By impacting local food production, water risks can cause different types of market and trade effects depending on the type of product, the amplitude of the shock and the market conditions. Assuming a significant shock in a larger market for a product with relative inelastic demand (and no immediate substitute), the relative prices will rise. The unabsorbed demand may be balanced by imports from non-affected producing regions. But even a significant shock in a smaller (price taker) market may result in relative entry of competing regions products that have acquired a cost advantage.

Market effects can be either short lived or result in long-term shifts. When extreme weather events hit regions that are significant exporters of agriculture commodities, international food prices can soar on spot and future markets. For example, the international price of wheat and maize rose by 25% in June and July 2012 after summer droughts in North America and Eastern Europe, which worried international markets and affected the global outlook for cereals and soya production (World Bank, 2012). On the other hand, gradual shifts in water risks that lead to production changes can result in long term market changes.

Market effects can also propagate internationally. Liu et al (2014) explored the impacts of irrigation risks on agriculture production and the role of international trade. Their simulations focus on regions that are expected to face irrigation failures, including China, South Asian countries, and the Middle East-North Africa region. The results show that a water decline would lead to significant production declines, shift irrigated land to other areas, and result in a significant reshuffling of international trade (Box 3.2). In particular, countries of Europe, North America, South America, and Oceania, as well as non-affected Asian countries are found to increase significantly their exports towards countries at risk of shortages.

Both market and trade effects can be amplified when water risks are prevalent in multiple regions at once. Links in the global water cycles can result in multiple impacts with wider effects. For instance, the 2002-03 drought impacted wheat production in Europe, the Russian Federation, Indian and China (Bailey et al., 2015). Bailey et al. (2015) argue that multiple shocks on wheat and rice production in Asia, Europe and the United States could lead to even greater response due to policy reactions (e.g. export restrictions), with large effects on global market, short-term and long-term food security.

Box 3.2. Simulating the effects of irrigation shortfall on cereal production and international trade

Liu et al. (2014) simulated the production and trade implications of a shortfall of irrigation by 2030 due to shift in demand in the absence of policy action. The projections on irrigation water availability from the IMPACT model are used to simulate production and bilateral trade effects using a current representation of the economy with the global computable general equilibrium model GTAP-BIO-W.

Figure 3.2 shows that the simulated reduction in water supply, which focused largely on China, South Asia and Eastern Mediterranean river basins, results in reduction of total crop output production ranging from 1% to 6%.The model also finds significant changes in land use. Regions under high stress reduce their irrigated land, and given productivity differences, extend their overall cropland, while regions not subject to these stress increase irrigated cropland. The overall pressure on land (+7.6 million ha) leads to reduced pastureland and forestry in many regions.

Figure 3.2. Effects of limited irrigation on water supply and total crop production

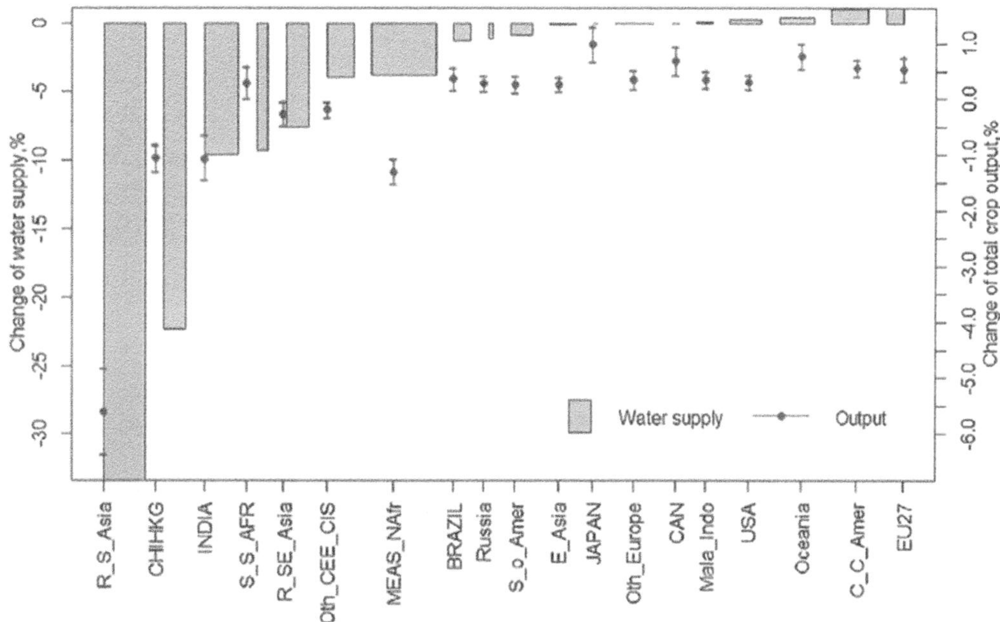

Note: R_S_Asia: Rest of South Asia, CHIHKG: China mainland and Hong Kong, China, S_S_AFR: Sub-Saharan Africa, R_SE_Asia: Rest of Southeast, Oth_CEE_CIS: Other CIS countries, MEAS_NAfr: Middle East and North Africa, S_o_ Amer: South America, E_ Asia: East Asia, Oth_Europe: Other Europe, CAN: Canada, C_C_ Amer: Central America. Bar width is proportional to the share of output value that is from irrigated crops in the region. Error bars associated with output indicate the 95% confidence interval of the mean for a normal distribution.
Source : Liu et al. (2014).

The study shows that these shifts in production translate into some redistribution of international trade flows. Price rises associated with higher costs of production in stressed regions alter the overall balance of net comparative advantages. As shown in Figure 3.3, affected regions (China, South Asia, and Middle-East North Africa)- significantly increase their imports from agriculture regions without these stresses, such as Europe, North America, the Rest of Asia, Latin America and Oceania. If China and India still export a little more they import much more from these regions. Overall trade increase in all commodities except sugar crops, dairy cattle, processed ruminant meats, processed rice, and processed food (where water-stressed countries tend to dominate the market).

(continues on the next page)

Figure 3.3. International trade shifts in maize, wheat and rice due to irrigation shortfalls

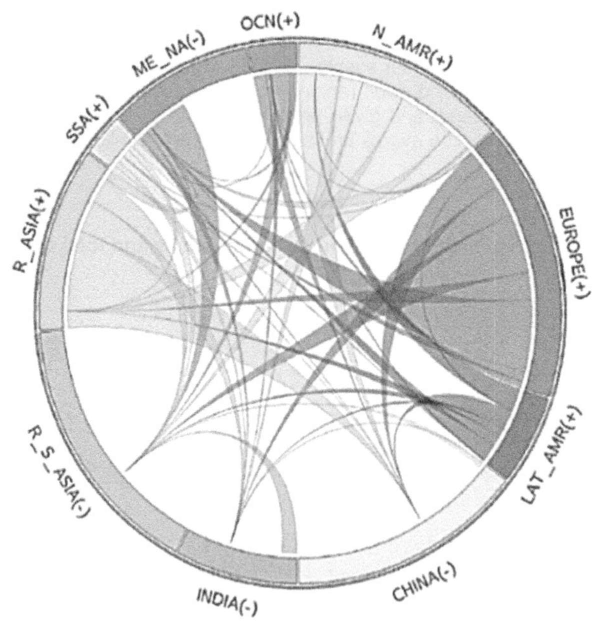

Notes: Wide sides of the arrow indicate the source of additional exports, narrow side indicates the importer; the size of the arrow represents the importance of trade shifts. The signs indicate net trade effect (+: export, - import). LAT_AMR: Latin America, N_AMR: North America, OCN: Oceania, ME_NA: Middle East and North Africa, SSA: Sub-Saharan Africa, R_S_Asia: Rest of South Asia, R_Asia: Rest of Asia.

Source : Liu et al. (2014).

These market and trade effects of water risks can impact food security in countries with large population of relatively low-income farmers or consumers. Local water driven agriculture shocks can affect poor and relatively market-isolated farmers, reducing their own staple food and revenue. At the same time, urban or market-connected consumers will suffer losses in income due to the increasing price of food. Willenbockel (2012) simulates the impact of extreme weather events on agriculture productivity market and food security in 2030 under a climate change scenario. A yield shock is implemented for the impacted country or region and the results are then measured for all regions. The results show for instance that a yield shock on cereals in North America, mimicking that in 1988, would result in world export price increases of 32% and 140% for wheat and maize respectively, impacting import and consumer prices in Sub-Saharan Africa. Similar effects are found for shocks on India and other East Asia, South America or Oceania, with different crops and amplitudes.

Sector impacts from water risks in the hotspot regions

In Northeast China, any impact on water shall have significant economic impacts in agriculture and beyond. Droughts, floods, storms and sea-level rise already generate large economic impacts in China, including in agriculture (Sadoff et al., 2015). The total loss due to water stress projected in 2030 in Northeast China could range from USD 1.1 to 1.7 billion (ECA, 2009).

Shifts in food demand due to a growing population and changing consumption patterns will likely increase water stress for agriculture. Though China's population may peak in 2030, the gap between food demand and supply was estimated to be around 68 million tons in 2040 (Yao, 2007). Shifts in consumption patterns may increase demand for water-intensive products, such as meat and dairy. Estimates for East Asia suggest that per capita consumption of meat will increase 38% by 2030 and 60% by 2050 (FAO, 2012). The projected increase in milk and dairy demand is even higher–48% by 2030 and 68% by 2050. Of course, the impact on water and trade depends on whether future demand is met by domestic production or imports. As livestock is very water intensive due to feed the decision to rely on domestic production would heavily intensify water stress if located in the Northeast. However, recent increases in meat and milk imports suggest trade may be at least part of the solution.

Collectively, changes in supply and demand may impact China's international trade positions. If agriculture productivity growth in the Northeast declines due to water scarcity, China may become less self-sufficient and increase imports. Although China has managed to deal with production shortages in the last years, increasing demand for food and feed will make it harder to rely on domestic production. In particular, further imports of water-sensitive grains, such as rice and corn, will be needed. Large stocks will only be able to buffer production gaps for a limited time. Assuming stable world food prices, this would increase China's already-rising import demand on global markets in the long run. Moreover, the rising trends in meat imports hint towards further reliance on global markets for meat and dairy products. At the same time, improving conditions for wheat and potato production could shift consumption patterns and exports of these crops.

Northwest India's agriculture sector impact may be significant because of the importance of the region in national supplies and exports of grains. A production drop in this region could significantly affect production and food security in India, since Punjab and Haryana supply 50% of the federal rice stock, and 85% of the federal wheat stock (Shiao et al., 2015). This food grain contributes to the national public distribution system through "fair price shops", providing subsidised food to poor households in the whole country, as required by the 2013 National Food Security Act. In 2013 approximately 800 Million people benefited from this policy, which represents 67% of Indian population (The Times of India, 2013). A cut in the main domestic rice provision source, or increased uncertainty in agriculture outputs of the Northwest, could also lead India to reduce its net exportations in order to support governmental stocks. India is currently the second top rice exporting country and will stay a major rice exporter, even if its world share is expected to decrease by 2024 (OECD-FAO, 2015). In the recent past, the volume and composition variations in India's rice stocks had enough significance to affect commodity prices at a global scale, as during the 2008 global food crisis (Mohanti, 2016; Childs and Kiawu, 2009).

In the Southwest United States, increased competition for more variable and uncertain water supplies will undoubtedly affect the agriculture sector. In addition to California's leading role in agriculture productivity and consumption, agriculture exports from the state have exceeded USD 18 billion annually since 2012 (CDFA, 2015). Water shortages and reallocation will affect production and exports, magnifying local impacts across the global stage (MacDonald, 2010). Field crops, livestock and dairy products, the most affected activities—due in part to a reallocation of water to high value specialty crops—represent a significant share of exports. In 2014, California's exports of dairy products, beef and related products, rice, cotton and hay amounted to USD 4.2 billion or 19.5% of the state's total agriculture export value (CDFA, 2015). Still, given the large number of variables affecting future agriculture productivity, including irrigation efficiency, crop shifting, improvements in crop yields, variability in global markets, and demand generally, it is extremely difficult to project the possible economic impacts of long-term drought and water reallocations beyond the region.

Local agriculture water risks can impact prices and trade: Illustration from a simulation of the international effects of water risk in the three hotspot regions

To complement the assessment of agriculture water risks impacts, a series of simulations of water risks is conducted in the three hotspot regions, using the International Model for Policy Analysis of Agricultural Commodities and Trade (IMPACT) version 3.1 (Robinson et al., 2015). The objective of the simulations is to assess the possible national and international market effects such risks might have in the future. Due to the difficulty of modelling a large range of water risks in each region, and the uncertainties they face, the exercise is designed as "water stress test" of the agriculture system rather than a set of accurate projections.[10]

The four main scenarios include a baseline with no water or climatic shocks (noted S0); the introduction of two types of gradual irrigation stresses in the three regions, aiming to capture (a) a reduction in usable surface water (mimicked by lowering basin efficiency, noted SB) and groundwater for agriculture (noted SG); and (b) exogenous yield shocks in particular years on key agriculture activities in the hotspot regions to reflect cyclical extreme weather events, and more specifically droughts (noted SD). These four scenarios (S0, SB, SG, SD) are then combined with the introduction of two global climate change simulations using regional climate projections RCP 8.5 IPSL S1 and Hadley S1. The full specification of scenarios is described in Annex 3.A1.2 and a background on extreme events used to design the scenario SD is provided in Annex 3.A1.3. Results are presented below as relative differences in production, international trade and prices between the different simulations and the baseline with no shock.

There are several caveats to the simulations. First, the selected hotspot regions are defined based on selected river basins, so they may not be exactly fit the previously defined hotspots. Second, the shocks are proxy water stresses rather than realistic water shocks, as per IMPACT 3.1 version, which does not include a full integration of hydrogeological systems with agriculture (see Annex 3.A1.2).[11] Third, each of the exogenous shock parameter was set up to stress the system in, but it may not reflect actual shocks. Fourth, because hydro climatic shocks are generally not independent, the scenario proposed do not account for water-related events that could happen in other regions of the world. Fifth, the two climate scenarios present two possible projections among other possible. The results of these projections should be subject to cautious interpretation, because of the lack of reliance of regional precipitation projections (Liu et al., 2014). Lastly, because IMPACT is a partial equilibrium model, price and trade effects are based on excess supply and demand equilibria at the global scale; they do not capture changes in bilateral trade relationships and they do not take into account of other sectors' feedback loops.

The main results are presented at the global scale, for hotspot countries and for trade partners for the two categories of risks: gradual water stresses and drought. Further results, including those combining some of the shocks with climate change projections, are shown in Annex 3.A2.

Gradual irrigation stresses in the three hotspot regions reduces field crop fruit and vegetable production in China, India, and in the US, increasing prices and affecting international trade in these products

The basin water deficiency (SB) and groundwater (SG) scenarios are run separately and in combination (SBG) simultaneously in the three hotspot regions and gradually implemented until 2050 without and with climate change. The results reported here pertain to the year 2050 and are outlined for key agriculture activities identified as possibly affected. Unless specified otherwise, the results are presented compared to the baseline S0 with no water or climate change shock.

Globally, irrigation stresses in the three regions decline global production in selected commodities compared to the baseline. Maize production is the first impacted with reductions of 3% to 4% (Figure 3.4, left). Cotton, wheat and temperate fruit production are also affected with declines over 1% of global production. Vegetables and rice are mostly affected under the combined scenario SBG. These losses are significantly amplified when adding the two climate change projections (Annex 3.A2). Maize production then decreases by close to 25% and 20% globally under the IPSL and Hadley projections, respectively. Cotton is the second activity affected with impact between 8% and 10%. Potato, rice and fruit follow with effects around 5-7%. Irrigation shocks only marginally affect these climatic losses; the SG scenario leads to lower effects than the SB scenario, and losses under the combined SBG scenario are greater than those with the individual scenarios.

Figure 3.4. Global production and price impacts with irrigation stresses

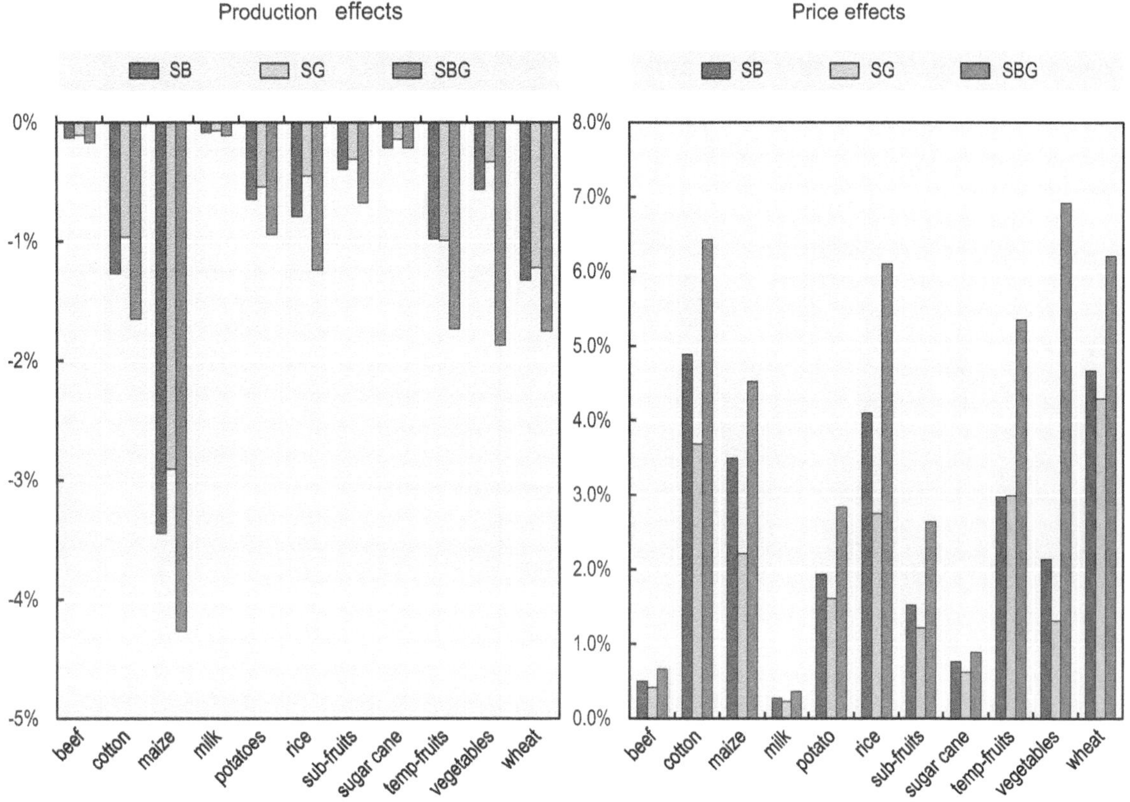

Note: Sub-fruits are subtropical fruits, temp-fruits: temperate fruits.
Source: Results from IMPACT simulations.

Consequently, global prices increase for directly concerned commodities (Figure 3.4, right), to a different degree. Heterogeneous price elasticities of excess demand for different products translated into a stronger price impact on wheat, cotton, rice and fruits than on maize. Overall, the price increases for these five activities range from 3% to 4.5% for the scenario SB, 2% to 4% for SG, and 4% to 7% for the combined shocks SGB compared to the baseline S0. As shown in the Annex 3.A2, climate change projections amplify absolute price effects significantly (over 10% increase for most activities). Maize, cotton and rice are the most affected with over 30% price increase under any scenario and projection. If these increases do not match the price impacts of recent food crises (1974, 2008, see Headey and Fan, 2011), they still would bear significance on the three countries and associated markets.

In China, the gradual deterioration of basin water efficiency and groundwater reserves in the Northeast and the two other hotspot regions affects the production and trade of cotton, maize, potato, rice, wheat, fruits and vegetables in 2050 (Figure 3.5). The largest negative impact is seen on maize and wheat, whose production drops by more than 10% under the scenario SG and the combined SBG scenarios. These effects result largely from yield losses of 8% to 10% depending on the scenario. The production of vegetables, fruits and rice also decline to a lower degree. These gradual stresses also result in a decline in the net trade in rice, fruits, and wheat (less than a 1%), and vegetables (up to 3% for the SBG scenario).

Figure 3.5. Relative changes in production and net trade in China by 2050 with gradual irrigation stresses

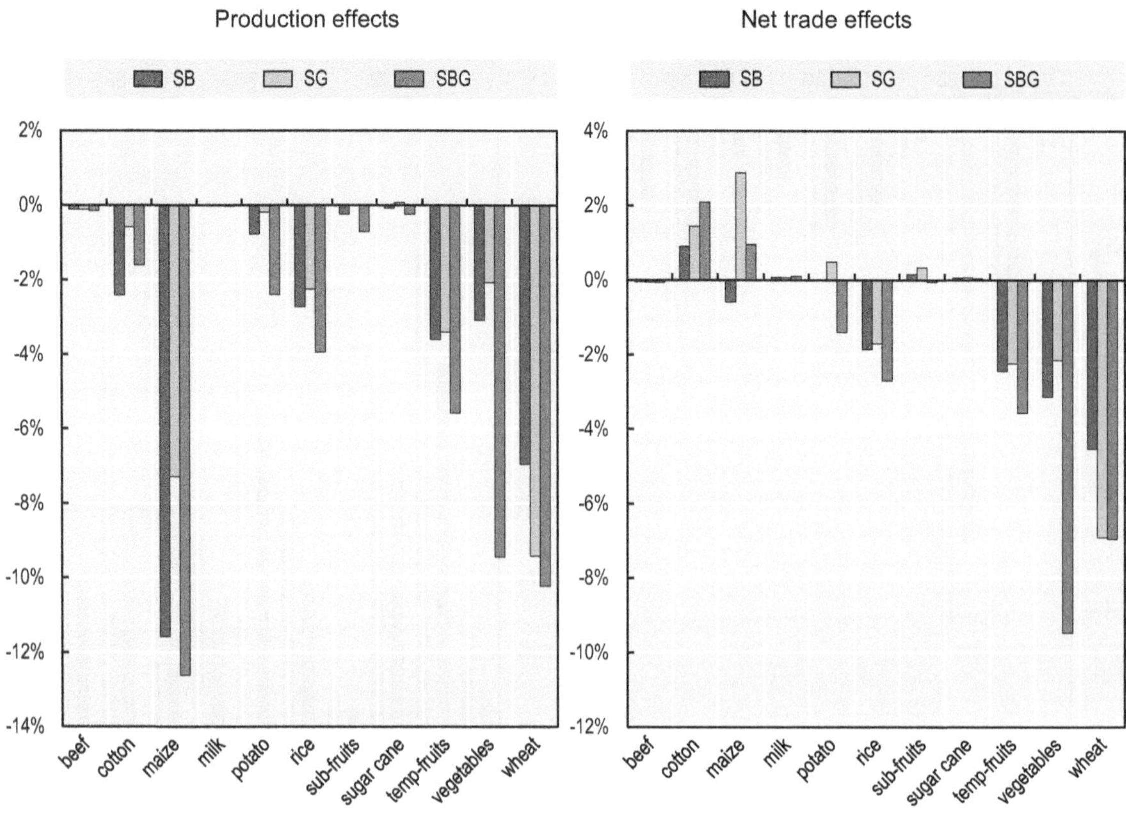

Note: Sub-fruits are subtropical fruits, temp-fruits: temperate fruits. Net trade changes may imply shifts in the direction of trade flows.
Source: Result from the IMPACT simulations.

Adding the two global climate change projections, however, alters the national production effects for most agriculture activities. Under the Hadley projection, the production of milk, potato, cotton and fruits in particular increases significantly—potentially due to temperature and precipitation changes in China and a change in relative competitiveness—but these effects do not prevent relatively large losses in the production of maize (by 16% to 22%, see Annex 3.A2). Under the IPSL scenario, maize production drops by 23% to 30%, while other changes remain small. Climate projections are generally more positive under the SG scenario than the SB one, with the SBG still being the most negative.

In India, gradual irrigation stresses in the Northwest region impact cotton, wheat and rice production, but these effects are lower than those observed in China and there is significant variation across the types of imposed irrigation stresses (Figure 3.6). This difference may be partly due to the fact that the Northwest (for rice or wheat) may not represent as large a national share of agriculture production as the Northeast China (for maize and wheat). Scenario SB reduces the production of cotton, rice and wheat by 2 to 4%, when the groundwater scenario reduces these productions much less (1% and even positive effect for wheat). These differences are due to the modelling differences -groundwater stresses lead to expanded areas for wheat and cotton at the national level and the respective competitiveness of the different crops in the three regions in this scenario. The three crops are most affected, however, under the combined scenario SBG. Under the same scenario, there is an increase in the production of vegetables responding to the significant production decline in China.

The effects on international trade only partially reflect these changes. Cotton and rice exports are reduced, but the maize trade balance is the most negatively affected. Following a small production increase, India also increases its trade balance for temperate fruits.

Figure 3.6. Changes in production and net trade in India by 2050 with gradual irrigation stresses

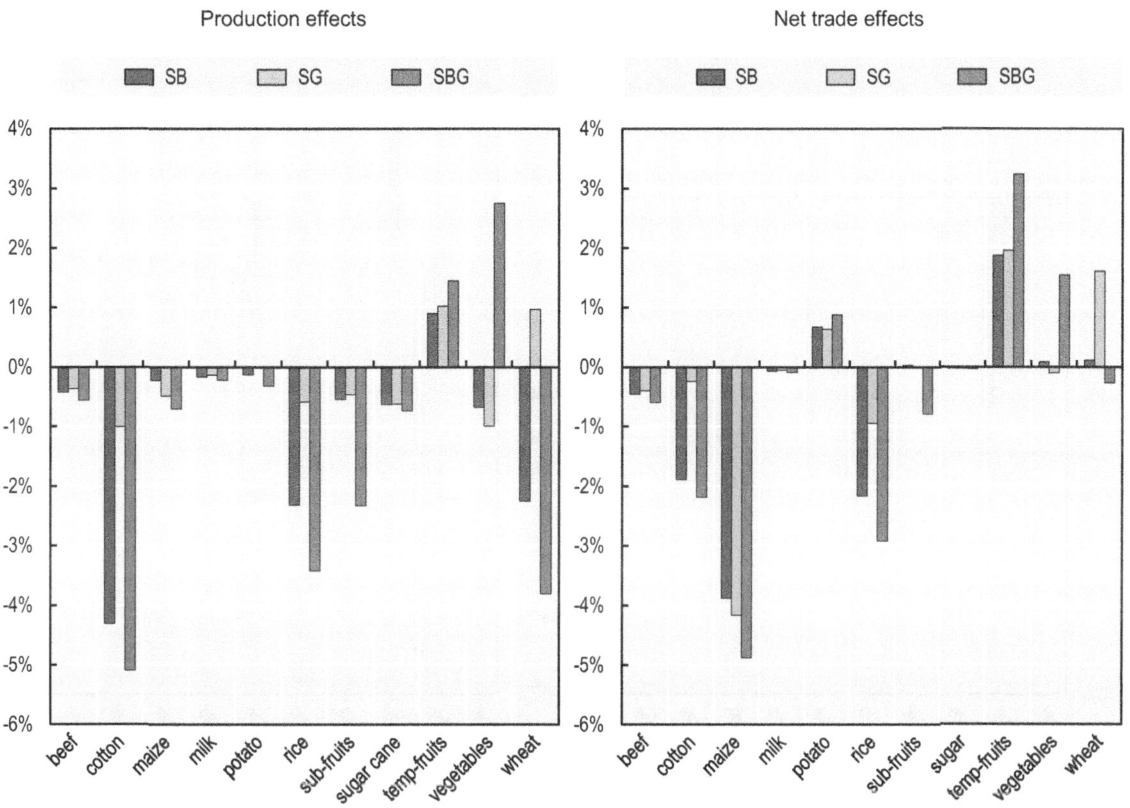

Production effects

Net trade effects

SB SG SBG

Note: Sub-fruits are subtropical fruits, temp-fruits: temperate fruits. Net trade changes may imply shifts in the direction of trade flows.
Source: derived from IMPACT simulations.

The results of the two climate change projections differ widely across projections (Annex 3.A2). The Hadley projection highlights losses of cereal production by over 20%. Wheat and vegetables production decline by over 25%, maize, rice and sugar cane by 20% or more, and cotton by 15%.These impacts marginalise the role of irrigation stresses. The IPSL projection, on the other hand, projects very small losses in the production of rice, wheat and vegetables, with a 10% production gain in maize production.

In the United States, irrigation stresses in the Southwest impact the production and trade of fruits, cotton, maize and rice. Production effects, however, only exceed 2% with the combined scenario SBG (Figure 3.7). In contrast with the other two countries, the groundwater scenario SG leads to relatively higher damages than SB, probably because the basin efficiency parameter starts at a much higher level than in China and India. Under SBG, fruit production, which is prevalently produced under irrigation in the Southwest (Cooley et al., 2016), declines by almost 9%. Exports drop by almost 11% for cotton, by 10% for rice and by 7.5% for fruits. Net trade for maize and wheat on the other hand increases slightly due to a relative gain in competitiveness compared to the other two shocked countries.

Climate change projections consistently lead to further production losses in maize, rice and cotton production under the three irrigation stress scenarios (Annex 3.A2). The amplitude of these effects is larger under the Hadley projection than the IPSL one. Under the IPSL projection there is an increase in potato production. The US net trade balance is reduced significantly in multiple activities, with significant losses especially for rice (-35%) and cotton (-20%).

Figure 3.7. Changes in production and net trade in the United States by 2050 with gradual irrigation stresses

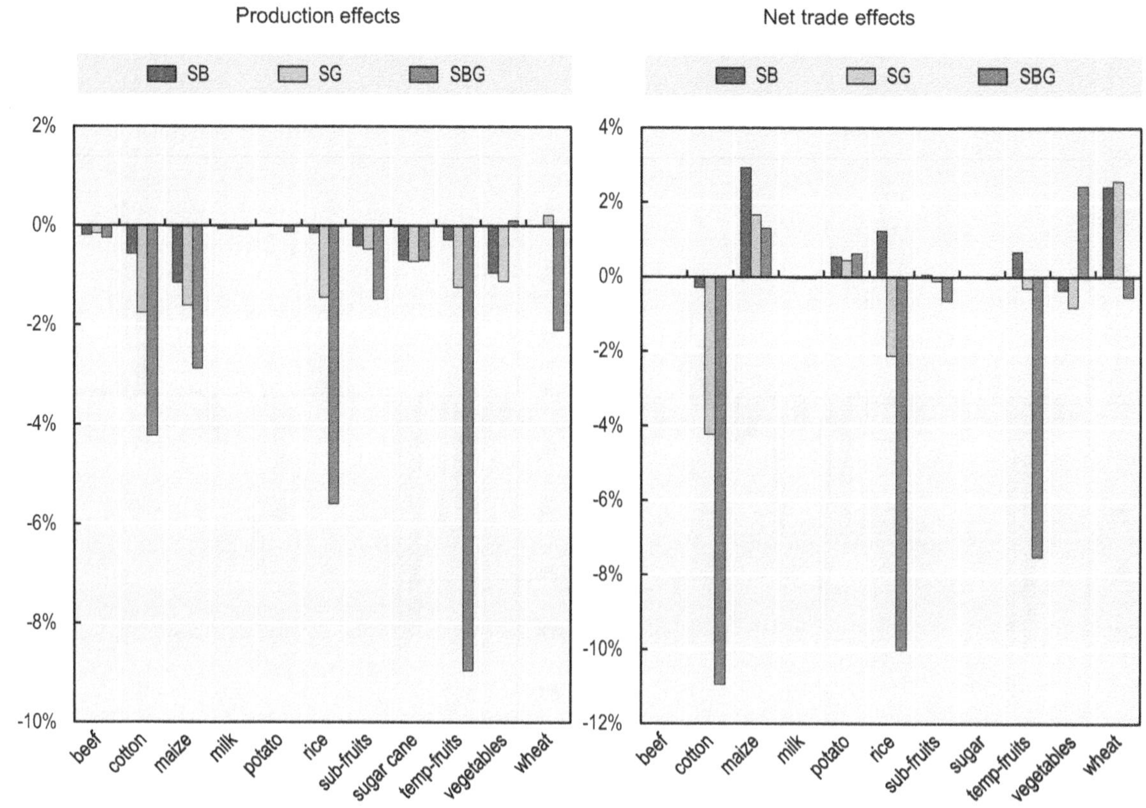

Note: Sub-fruits are subtropical fruits, temp-fruits: temperate fruits. Net trade changes may imply shifts in the direction of trade flows.

Source: Derived from IMPACT simulation results.

The price impacts also trigger responses in international markets, leading to changes in net trade balances across countries, particularly for cotton, maize, rice, and wheat. The main impacts of the different scenarios, i.e. irrigation stresses with and without the two climate projects, are reported in Table 3.4 for the most significant relative trade effects for the different commodities. The complete table of effects is presented in Annex 3.A2. Major changes are foreseen in particular in the cotton, maize, rice and wheat markets, consistent with the most significant production losses observed in the three hotspot countries. Major agriculture exporters, such as Australia, Brazil, Chile, or European nations, gain in global market shares. In contrast, beef and dairy markets remain largely unaffected.

Table 3.4. Largest relative net trade effects of the irrigation stresses and climate change scenarios by commodity[1]

Commodity	Scenarios	Most important net trade effects	Countries
Cotton	All except Hadley projections	+49% to +194%	Australia-NZ
	All scenarios	+1% to +21%	China
	All scenarios except SB	-11% to -43%	United States
Maize	All scenarios	+1% to +33%	Australia-NZ
	All scenarios	+1% to+ 25%	Brazil and Chile
	All scenarios	+1% to +22%	EG4
	All scenarios	-4% to -7%	India
	All scenarios	-8% to -23%	Japan
	All scenarios	-9% to -81%	Russia
Rice	All scenarios	+9% to +15%	Brazil
	All scenarios	+3% to +37%	E17
	All scenarios	+1% to +34%	Chile
	All scenarios	-1% to -15%	India
	SBG scenarios	-10% to -66%	United States
Wheat	All scenarios	+3% to+ 39%	Russia
	SB, SG, SBG scenarios	+4 to +5%	EG4
	All scenarios	-3% to -18%	Australia-NZ
	All scenarios	-5% to -22%	EU-7
	SG scenarios	+2 to +6%	India
	SB, SG, SBG scenarios	-5% to -7%	China
	All scenarios	-7% to -97%	Brazil
	SB, SG, and SBG (no climate projection)	+12%	EU-7
	All scenarios	-30% to -43%	E-17
Vegetables	SGB scenario	+4% to +31%	Chile
	SGB scenario	+3% to +16%	Japan
	All scenarios	-2% to -12%	Russia
	SB, SG, and SGB (no climate projection)	-13% to -17%	Canada
	All scenarios	+4% to +58%	Chile
Temperate fruits	All scenarios	+2% to +7%	India
	SGB scenarios	-7.5 to -11%	United States
Sugarcane	All scenarios	+1 to +11%	Brazil and Chile

Note: 1. The absolute size of trade change depends on the initial volume of trade flows, net trade changes can result in shift in the direction of trade flows, which should be accounted for when interpreting these results. EG4: France, Germany, Italy, United Kingdom, E17: Austria, Belgium, Luxemburg, Netherlands, Ireland, Spain, Portugal, Greece, Czech Republic, Slovakia, Poland, Hungary, Estonia, Finland, Denmark, Sweden Slovenia; EU7 : Bulgaria, Lithuania, Latvia, Croatia, Malta, Cyprus, Romania.

Note by Turkey: The information in this document with reference to "Cyprus" relates to the southern part of the Island. There is no single authority representing both Turkish and Greek Cypriot people on the Island. Turkey recognises the Turkish Republic of Northern Cyprus (TRNC). Until a lasting and equitable solution is found within the context of the United Nations, Turkey shall preserve its position concerning the "Cyprus issue".

Note by all the European Union Member States of the OECD and the European Union: The Republic of Cyprus is recognised by all members of the United Nations with the exception of Turkey. The information in this document relates to the area under the effective control of the Government of the Republic of Cyprus.

Source: Derived from IMPACT simulations.

These simulations show that irrigation stresses are bound to affect field crop activity first and foremost; furthermore, maize and wheat are projected to be more impacted than rice in the three regions. At the same time, cotton and temperate fruit, water-dependent crops that are largely produced in the three countries, also face significant production losses. Basin deficiency scenarios lead to greater impact than those representing groundwater depletion, except in the United States. Climate change projections alter the impacts especially in China and India, although both still face losses in field crops. International prices increase by 2% to 6% for cereals, and trade is reshuffled, especially for field crop commodities–cotton, maize, rice and wheat- that are largely produced in the three hotspot regions.

Drought scenarios: Limited impacts of a drought in the US Southwest, measurable field crop production and trade implications of a combined drought in Northeastern China and Northwestern India

This scenario investigates the impact of extreme events, in particular due to changes in precipitation cycles. A rapid assessment of the frequency, duration and impacts of El Nino Oscillations (ENSO), Pacific Decodal Oscillations (PDO) and Indian Ocean Dipole (IOD) in the three hotspot regions was conducted, based on past and recent trends (see Annex 3.A1.3). This assessment was used to provide plausible types of events affecting the hotspot regions in the future. Two sub-scenarios are run: scenario SD1, modelling a drought in the US Southwest around 2021, and scenario SD2 whereby China and especially India are subject to drought around 2030.

Given the lack of adequate modelling tools to assess the full impact of droughts, exogenous yield shocks are modelled in each of the concerned region for key agriculture activities (as explained in Annex 3.A1). It thereby replicates the exercise conducted by Willenbockel (2012). The main results are shown below, with some complements in Annex 3.A2.

Under scenario SD1, a drought in the Southwest United States in 2021 induces significant reduction of yields mostly of field crops (Annex 3.A1.2). The drought is insufficient to induce significant impacts on global markets (Figure 3.8 right). Global production impacts are significant only when combining the drought with climate projection, resulting in production declines for maize, cotton and rice (Figure 3.8 left). World prices for vegetables increase by 2% under the drought, and much more for maize, cotton and rice under climate projections (over 8%).

Figure 3.8. Changes in global production and world prices with a drought in 2021 in the US Southwest

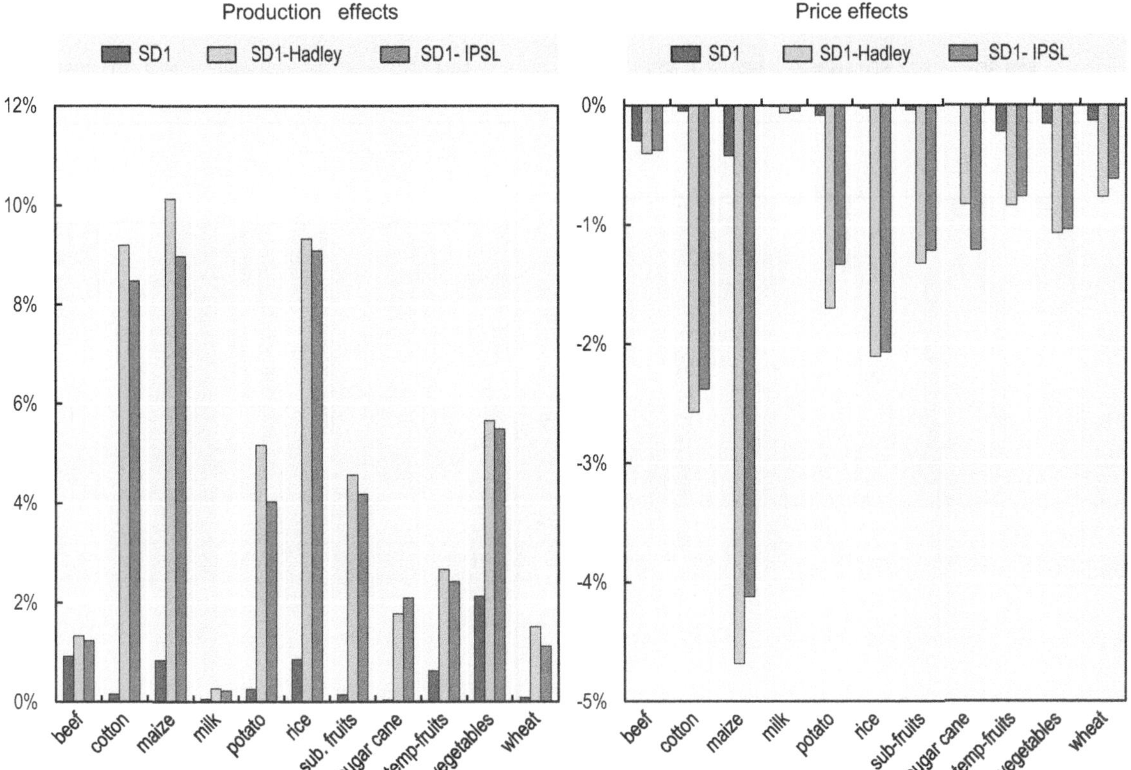

Note: Sub-fruits are subtropical fruits, temp-fruits: temperate fruits. Net trade changes may imply shifts in the direction of trade flows.
Source: Derived from IMPACT results.

The imposed yield shock in the Southwest United States leads to a reduction of production of the relevant activities in the United States, particularly vegetables and fruits (-8% and -6%, respectively), given the importance of the region to their national production (Figure 3.9).[12] Beef production is also reduced by 3%. Similarly, this leads to significant reduction of trade in fruits and vegetables, and to a lesser extent beef. The trade balance of wheat is also reduced by 4%. The drought production effects are stronger than that of climate change for the main affected crops.

Figure 3.9. Changes in production and trade in the United States with a drought in 2021

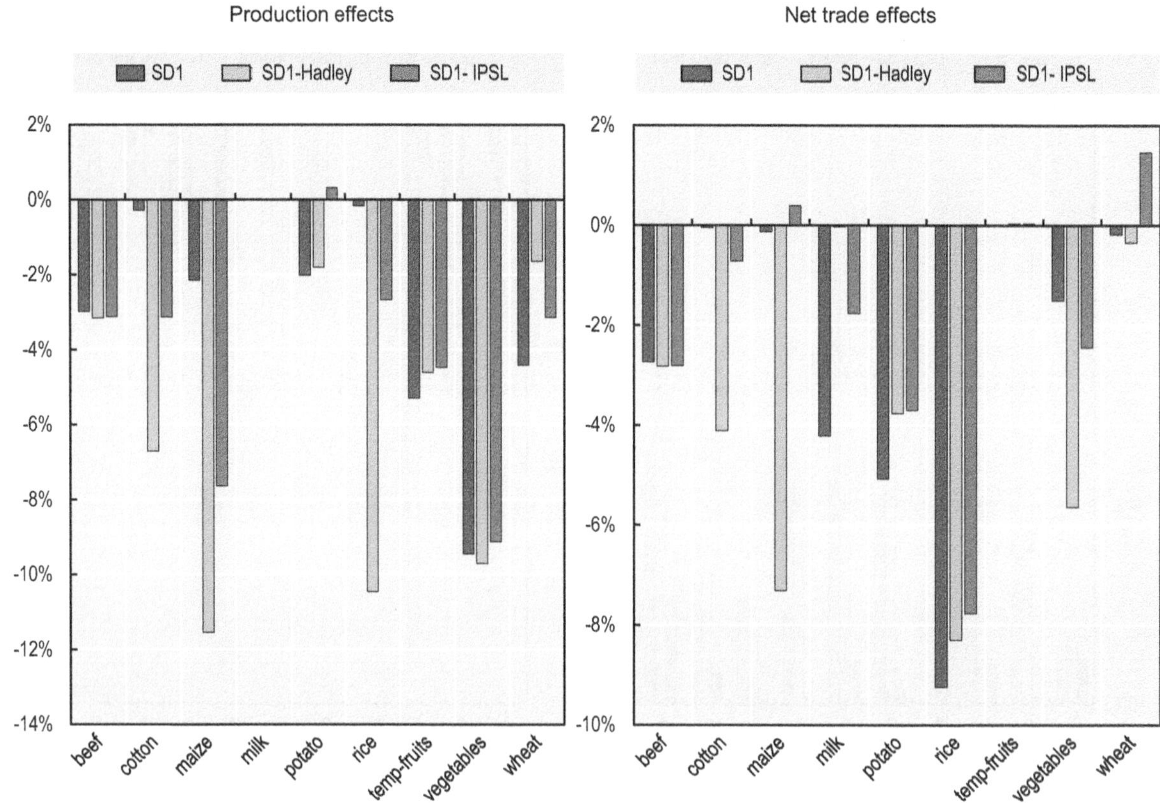

Note: Temp-fruits are temperate fruits. Net trade changes may imply shifts in the direction of trade flows.
Source: Derived from IMPACT results.

International effects from the SD1 scenario are limited. In particular, the 2021 Southwest United States drought affects agriculture production in China and India marginally relative to the baseline or climate projections (Annex 3.A2). Other countries gain small increments in market shares and limited reshuffling derived from price increases. Naturally, adding global climate change projections induces much larger changes in trade.

A drought in China and India in 2030, designed to reduce yields of several crops (Annex 3.A1.2) also translates into production losses of field crops. Losses of 2% to 3% of world production for wheat, cotton and maize (Figure 3.10) trigger price increases by over 5% for these crops as well as rice, vegetables and potatoes. These impacts are amplified with climate projections; e.g. price increases over 20% for cotton, rice and maize.

Figure 3.10. Changes in world production and prices with a drought in NE China and NW India in 2030

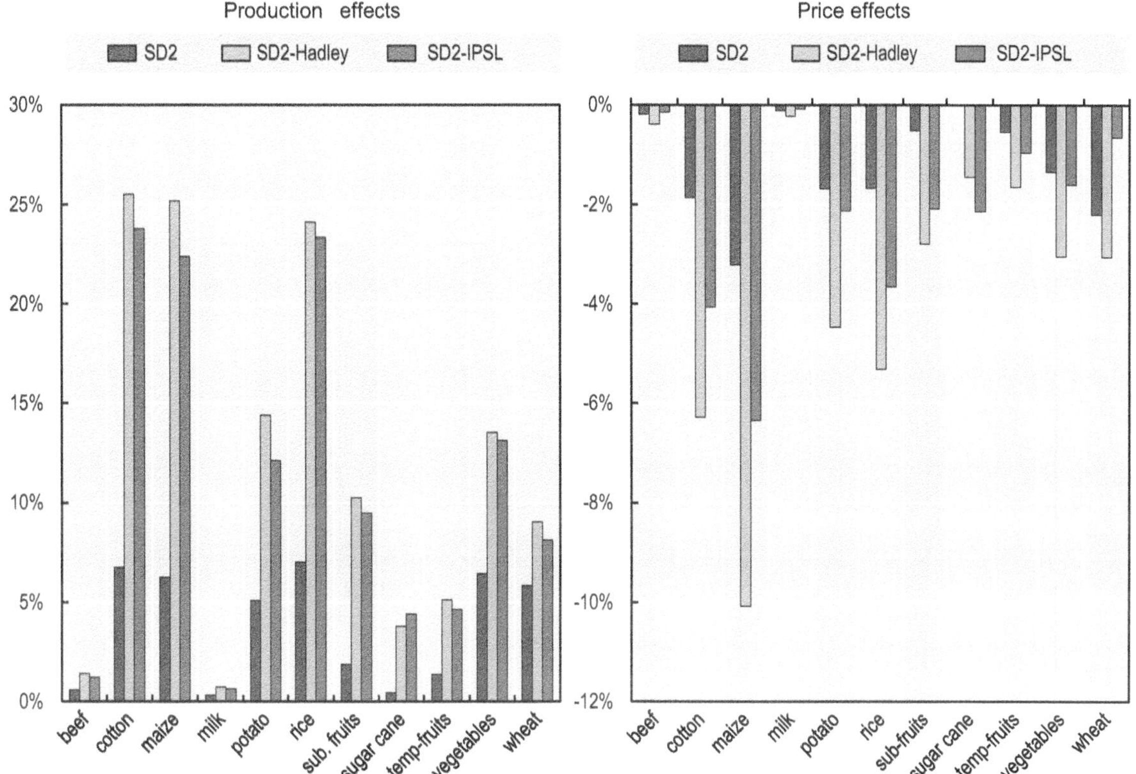

Note: Sub-fruits are subtropical fruits, temp-fruits: temperate fruits. Net trade changes may imply shifts in the direction of trade flows.
Source: Derived from IMPACT results.

As modelled, the drought in Northeast China does result in significant losses in production of maize and wheat, cotton and potato (exceeding for instance the 2.5% grain production reduction of the 2002-03 drought in China see Annex 3.A1.3) (Figure 3.11). These impacts induce a reduced trade balance in maize, potatoes and wheat. In both cases, the drought impacts are similar to that of climate change projections.

Figure 3.11. Changes in production and trade in China with a drought in Northeast China and Northwest India in 2030

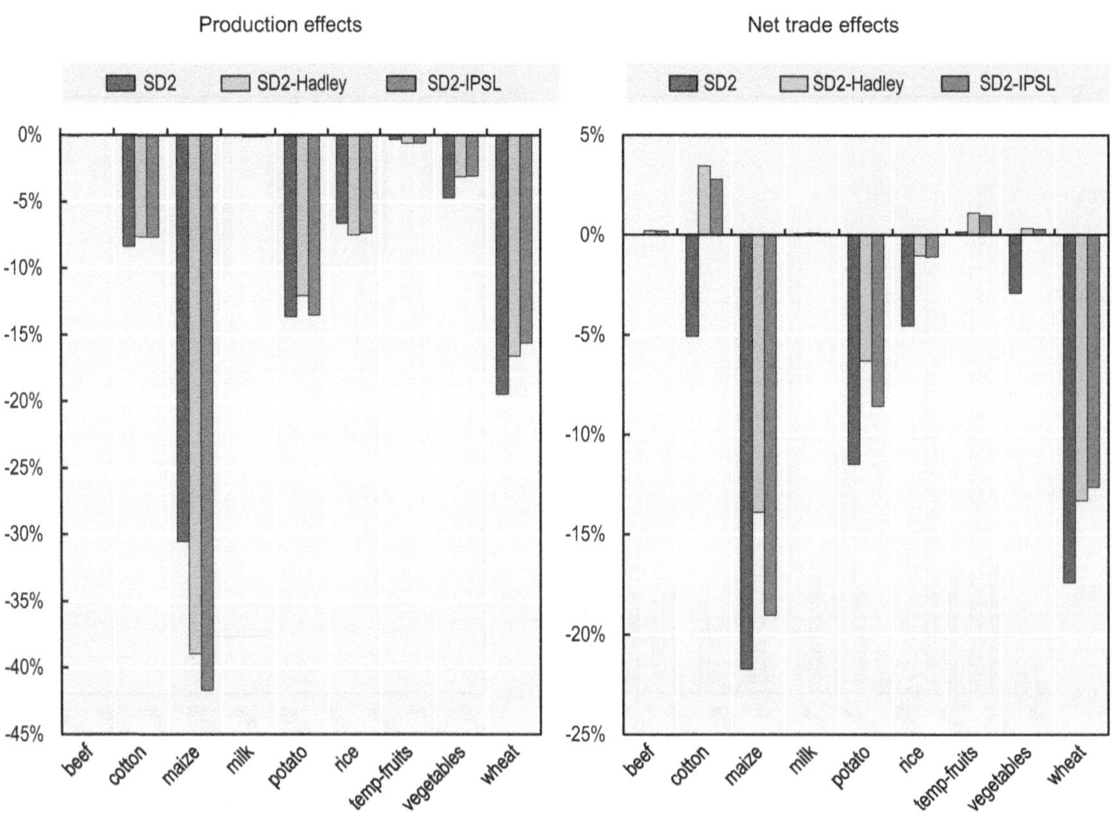

Note: Temp-fruits are temperate fruits. Net trade changes may imply shifts in the direction of trade flows.
Source: Derived from IMPACT results.

In contrast, India is still more impacted by climate change than by the drought (Figure 3.12). Wheat and rice are the most affected by drought and climate change. The impact on wheat is similar to that reached in the 2015 drought (Annex 3.A1.3). Cotton and vegetable production increase under the drought and decline significantly under the climate change projections. These results are directly translated into net trade effects.

Figure 3.12. Changes in production and trade in India with a drought in NE China and NW India in 2030

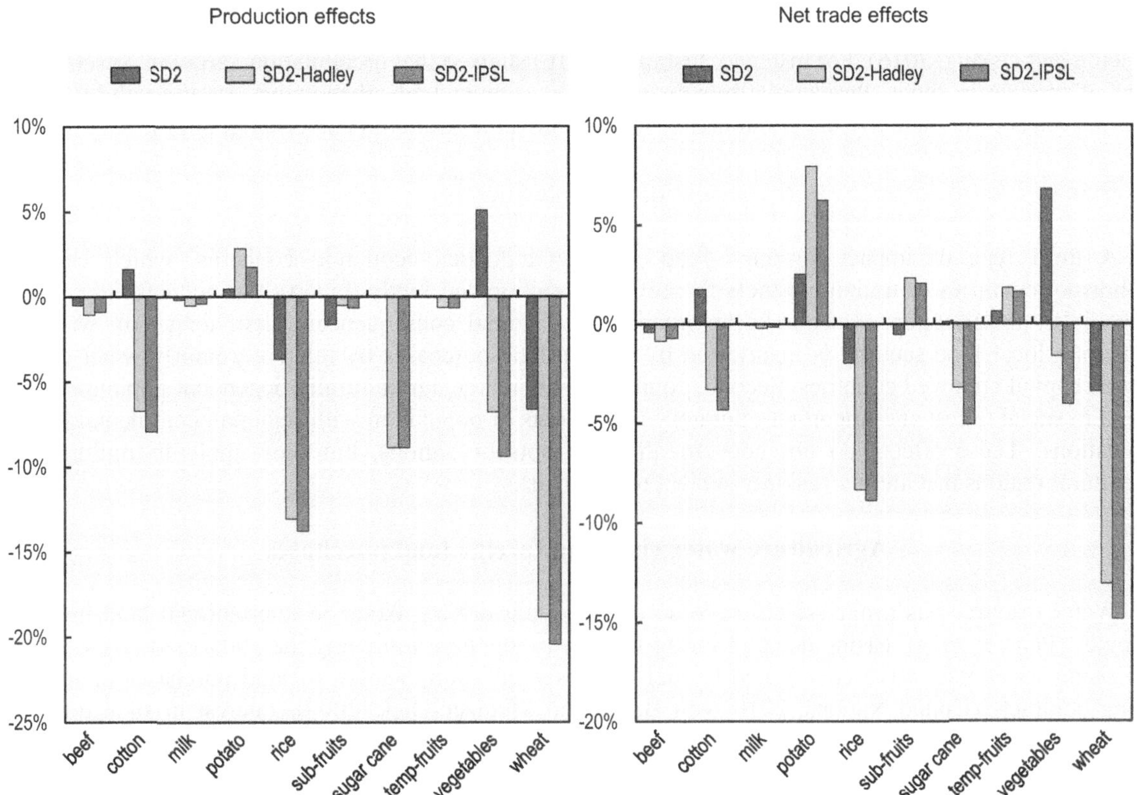

Note: Sub-fruits are subtropical fruits, temp-fruits: temperate fruits. Net trade changes may imply shifts in the direction of trade flows.
Source: Derived from IMPACT results.

The drought in China and India in 2030 affects other countries' trade balance more than the 2021 drought in the Southwest United States (Annex 3.A2). Significant trade effects are found in large countries especially for the most impacted commodities. The reduction of the trade balance for cotton in China results in increased of cotton exports by other countries, including Brazil (+6%), and Australia (+143%). Significant changes are also found in the maize market with increase in exports from Brazil and from the United States (+7%). Net trade increases are spread out across countries for rice, potatoes and vegetables. The lowering of the trade balance of wheat in China and India also creates room for competitors, with an increase in net trade by 6% to 7% in Europe, the Russian Federation, Canada and the United States.

The results of this exercise demonstrate that weather shocks in key agriculture regions can lead to significant temporary shocks on the market of water dependent crops, but that the amplitude and distribution of impacts will clearly depend on the region affected. The shock in the Southwest United States is largely absorbed internally, while shocks in Northwest India and Northeast China can trigger market and trade implications. Market volatility will clearly increase faster with shocks in large production areas for crops that are subject to water risks. At the same time, these simulations indicate that production shocks may be foreshadowed by the impact of climate change in vulnerable areas. Unlike the simulated shocks, which were implemented in three regions, climate change affects all countries and regions, so the effects on competitiveness are likely to be different. Furthermore, because climate change affects not only water but also temperature it is projected to have stronger effects at least in some countries.

Nevertheless, the literature remains undecided as to whether gradual changes versus extreme shocks will be most significant. Empirical studies using past data have shown that gradual warming could have less impact than inter-seasonal variation and weather events, at least in some locations (Carleton and Hsiang, 2016). For instance, Fishman (2016) showed that precipitation variation offsets the heat effect of models in India. Projections remain unable to capture both phenomena, so the validity of their projected impacts, particularly in agriculture, is still questionable (Carleton and Hsiang, 2016).

3.4. Broader food security and socio-economic concerns

A third type of impact regroups food security and socio-economic concerns, which can affect non-hotspot countries or trading partners thereof. This section succinctly reviews two increasingly observed pathways from agriculture water risks to broader international consequences. First, long-term water risks threatening local food security is manifested by foreign land purchases by relatively poorly water endowed but well capital endowed countries. Second, countries with a large agriculturally dependent population can be subject to social unrest and important migratory movements of population with regional, continental or global implications. These effects do not concern all three hotspot regions, but they apply to multiple other agricultural regions that are increasingly subject to water risks.

Agriculture water risks and foreign land purchases

Water resource has progressively become an important driving factor in international land investment strategies (Pearce, 2013). Rising demand for agriculture production, following the 2008 food crisis, and long term rising water risks have contributed to fuel the observed multiplication of land purchases or long-term leasing contracts (United Nations, 2010; von Braun and Meinzen-Dick, 2009). Interest in such transaction stems from unequal land and water endowment, originating especially from water-poor countries,[13] such as Gulf States, but also from countries with large population such as China or India concerned with broader food security concerns by (Ibid.), and from private companies in Europe, North America, South Korea or Japan (Smaller and Mann, 2009). These investments take place on a wide range of countries; most transactions focus on developing countries especially in Africa, but some are also taking place in developed countries (Smaller and Mann, 2009).

Foreign land purchase is mostly undertaken by private companies, such as agribusiness, and investment banks that anticipate a growing value for access to fertile land linked with freshwater, with the purpose of local consumption or exports. At the same time, governments are often involved either directly or through state-owned entities, taking an active part in negotiations, with the objective of responding to their own internal food demand (von Braun and Meinzen-Dick, 2009). For instance, in 2008, Saudi Arabia shifted its strategy from mining its groundwater for agriculture irrigation to investing into water-endowed foreign land for food and feed procurement by establishing the "King Abdullah initiative for Saudi agriculture investment abroad" (Smaller and Mann, 2009).

If some of these transactions may provide opportunities and can be considered as a specific type of foreign direct investment, many others raise issues about property and water rights, leading to tensions or even conflicts within local populations (Zetland and Möller-Gulland, 2012). Some perfectly legal transactions may potentially worsen long-term water risks in hotspot areas (see Box 3.3). In other cases, they may threaten long-term food security by giving land and water rights away, often without assessing the current and future consequences for agriculture or water users (Skinner and Cotula, 2011; United Nations, 2010; Woodhouse, 2012). These deals also can generate broader political conflicts, as observed in Madagascar (von Braun and Meinzen-Dick, 2009).

Box 3.3. Investing in irrigated land in a productive hotspot region: Producing hay in the Southwest United States

Several recent land investments were made by large Gulf State companies in the Southwest United States. In particular, Almarai, the largest dairy company in Saudi Arabia, bought 5666 hectares (14 000 acres) in California's Palo Verde Valley and in Arizona to grow and export back alfalfa for cattle feed.

Both acquisitions were explicitly pursued to alleviate agriculture water shortage risks. Alfalfa is a water intensive crop whose cultivation has been banned temporarily in Saudi Arabia because of water shortage, prompting Almarai to look for foreign land with more secure water. Despite the Southwest facing its own water challenges, the company ensured that it could benefit from its productivity while keeping a steady flow of water. To do so it purchased land in with senior water rights in California, with priority over the Colorado River Basin. And in Vicksburg, Arizona, it purchased quasi deserted land that could be irrigated with groundwater given the lack of clear regulatory oversight.

If these perfectly legal transactions have resulted in farmer engagements and approvals, creating business opportunities, it also created concerns from local water users in Arizona, whereby aquifer levels are rapidly dropping. In the longer term, it does raise the question of the sustainability of the system, and possible raising conflicts in the future, especially when considering the growing water risk shortages in the region.

Source: NPR (2015); Spagat and Batrawy (2016).

Since the land contracts are signed under international law, they are often more binding to the states than the ownership of local communities under domestic law, whose rights may be unclear outside of the traditional context (Smaller and Mann, 2009). Therefore, in case of dispute, the local owners or users may have difficulties to claim their rights against expropriation, notably regarding access to water resources. As a consequence, foreign agriculture investment is often associated with decreased standards of living and loss of subsistence means for local communities and long term consequences with water users. On the other hand, international law guarantees to foreign investors the access to the necessary means of production, including access to local freshwater.

The number of such land purchase is substantial but remains difficult to monitor and assess; still observers believe that it could increase in the future along with water and climate risks. Overall, it was estimated that appropriation of water rights associated with land purchases represent 140 billion m3 of surface and groundwater each year (Rulli et al., 2013). In Africa, large-scale land acquisition covered 22 million ha (the size of the United Kingdom); of these only 3% were found to be used for production in 2015, but the production mostly focused on relatively high water requirement crops (sugarcane, jatropha, eucalyptus) (Johansson et al., 2016). Hundreds of examples are discussed in the literature, even if they are not always well documented (von Braun and Meinzen-Dick, 2009; Zetland and Möller-Gulland, 2012). It is likely that the trend will continue to increase with climate change and growing water scarcity and food security tensions (Smaller and Mann, 2009; United Nations, 2010).

Agriculture water risks, conflicts and migration

There is a growing concern that water risks impacts on food security have fuelled conflicts and migration. In January 2017, agriculture ministers of the G20 countries signed a declaration in which they note that "Water scarcity and excess water threaten agriculture and food security and nutrition. This can contribute to political and social instability and to large –scale migration" (G20, 2017). As noted in Chapter 1, the World Economic Forum has ranked water crises as a major risk factor (WEF, 2015, 2016 and 2017). The importance of security risks has also been identified by intelligence agencies (Office of the Director of National Intelligence, 2012, cited in Siegel, 2015).

A growing body of research has assessed the potential links between climate change, water availability and conflict. If there is mixed empirical evidence on the impact of droughts and temperature shocks and conflicts globally (Gleditsh, 2012; Nillesen and Bulte, 2014), such link has been substantiated in a number of specific cases, especially in the African context. Civil wars in Somalia, Sudan, and South Sudan were largely fuelled by drought (Breisinger et al., 2015). Other conflicts have been partially attributed to drought and climate change (Gleick, 2014; Kelley et al., 201; Jägerskog et al., 2016).

In most of these cases, a climatic event was associated with ongoing water stress, creating production shortages and income disruption, deepening inequalities and exacerbating existing social grievances (Breisinger et al., 2015). Conflicts arose in rural areas, often moving to urban areas then lifted into tensions with governmental authorities, due to insufficient responses. The implications range from temporary political crises to outright civil or international wars.

An indirect implication of tensions growing importance of water risks relates to migration. If climate events have long been a driver of migration (Jägerskog et al., 2016), the relationship between climatic events and migration is also empirically ambiguous, due to its relationship with multiple socio-economic variables (Black et al., 2012). Still, a migration-climate risk link was established more convincingly in some regions and circumstances. Sea level rise, storms, floods, cyclones, or droughts have encouraged people to migrate in many countries, although they also were fuelled by other socio-economic and political factors, including an inadequate management of water and land (Black et al., 2012; Weiss, 2015).

Recent studies have focused on the possible link of worsening climatic conditions associated with climate change in certain areas and migration, which could then induce conflict. Given the large uncertainties, estimates of climate- induced migrants vary widely, from 25 million to 1 billion by 2050 (Weiss, 2015). Werz and Conley (2012) identified four regions most likely to be subject to intense migration in the future, due to climate change and water insecurity: Northwest Africa (international migration due to climate change, food and water insecurity), Bangladesh (floods and increased sea-level) and India (population growth), the Andean region (melting glaciers), and China (continued internal migration due to climate change).

If the reported links between water risks, especially affecting agriculture, conflict and migration remain complex and would warrant further research (Post, et al., 2016), the evidence on increasing water risks suggest that these links may strengthen in the future in the absence of policy response.

Notes

1. These indirect market effects also include possible trade policy responses derived from market shocks induced by water risks in hotspot countries.

2. Damages of droughts and floods (FAO, 2015) refers to agriculture losses as "the changes in the economic flows arising from the disaster": decline in crop, livestock, aquaculture output; increased costs of farm inputs and services; changes to food market prices

3. The remaining share of damages were incurred by forests, fisheries and irrigation, acknowledging that the forest and fisheries impacts were likely to be under-reported (FAO, 2015)

4. Not all extreme event result in long-term effects. With a large pool of events viewed from on a global scale, Lesk et al. (2016) found no significant impact of extreme weather disasters on crop production in the long term: production resumed its pre-disaster level the following year.

5. Cross-study comparability is limited as the results are largely dependent on assumptions, such whether or not farms have access to sufficient irrigation and the effects of CO2 fertilization (Piao et al., 2010; Wang et al., 2014).

6. The amplitude of the effects of climate change on irrigation could however be limited; Döll (2002) suggests that net irrigation requirements would increase less than the regular inter-annual variability.

7. Groundwater used beyond recharge.

8. Bad water quality – compounded by rising water scarcity - may also trigger negative health impacts. Water pollution in the Yellow River watershed has already created excessive levels of chromium and lead in rice and cadmium in cabbage. High rates of mental and physical development challenges have also been attributed to arsenic and lead contamination of food and water along the Yellow river. In rural areas the mortality rate for diseases related to water pollution is the highest, especially for stomach cancer and liver cancer and is far above world average. Given the projections of increasing pollution, these effects are likely to worsen in the future.

9. Shallow groundwater depletion creates incentives to drill deeper tubewells, up to 200m depth, which induce vertical leakages, compromising the recharge of shallow aquifers and thereby rising pumping costs even faster (BGS, 2015).

10. More countries or regions could have been subject to shocks, to reflect the determination of future water risks (chapter 2), but it would not serve the purpose of evaluating the impact of water risks in a few selected regions.

11. Irrigation water demand and supply are not formally defined in the IMPACT model, as they are exogenous variables coming from the Impact Water Model Simulation Model (IWSM).

12. It may be that a drought-induced water-supply shortfall would be reflected primarily in a reduction in acreage and less in terms of a reduction in yield. But perhaps it is more likely to see yield reductions for perennial fruit/nut tree crops where farms are more dependent on surface water supplies. The IMPACT model version used for this report does not allow running simulations for

perennial fruits and vegetables specifically as they are aggregated into the "fruit and vegetables" category.

13. At a global scale, water gaps are considerable: countries of the Gulf Cooperation Council use 80% of their water for irrigation, while it is only the case for 2% of the water supply in Africa, revealing largely untapped resources.

References

Agweb (2016), "Farmers in China Wrestle With Drought", Associated Press, 3 August 2016. www.agweb.com/article/farmers-in-china-wrestle-with-drought

Alexander, M. A. et al. (2008), "Forecasting Pacific SSTs: Linear Inverse Model Predictions of the PDO", *Journal of Climate*, Volume 21/2. https://doi.org/10.1175/2007JCLI1849.1

Andrews E. D. et al. (2004), "Influence of ENSO on Flood Frequency along the California Coast", *Journal of Climate*, Volume 17/2. https://doi.org/10.1175/1520-0442(2004)017<0337:IOEOFF>2.0.CO;2

Bailey, R. et al. (2015), "Extreme weather and resilience of the global food system", Final Project Report from the UK-US Taskforce on Extreme Weather and Global Food System Resilience, The Global Food Security Programme, London.

Bhatnagar, G-V. (2015), "Punjab seeks relief for crop damage", *The Hindu*, India, 27 March. www.thehindu.com/news/national/other-states/punjab-seeks-relief-for-crop-damage/article7037672.ece

Black, R., et al. (2012), "Migration, immobility and displacement outcomes following extreme events", *Environmental Science and Policy*, Vol. 27, pp. S32–S43. DOI:10.1016/j.envsci.2012.09.001

Bowers J.-C. (2001), "Floods in Cuyama Valley, California, February 1998", U.S. Geological Survey Water Fact Sheet 162-00. http://pubs.usgs.gov/fs/fs-162-00/

BGS (British Geological Survey) (2015), ''Groundwater resources in the Indo-Gangetic Basin - Resilience to climate change and abstraction'', British Geological Survey Open Report, Natural Environment Research Council, 31 July 2015.

Breisinger, C. et al. (2015), "Conflicts and Food Insecurity: How Do We Break the Links?" in International Food Policy Research Institute, *2014-15 Global Food Policy Report*, IFPRI, Washington DC.

Brown, M.E. et al. (2015) "Climate Change, Global Food Security, and the U.S. Food System", U.S. Department of Agriculture, the University Corporation for Atmospheric Research, and the National Center for Atmospheric Research, USDA, Washington, DC. http//www.usda.gov/oce/climate_change/FoodSecurity2015Assessment/FullAssessment.pdf

Browning, E. (2013), "Floods and Droughts Causing Major Crop Problems in China", *Financial Sense*, 11 September. www.financialsense.com/contributors/evelyn-browning-garris/floods-droughts-crop-problems-china

Bruinsma, J. (2003), ''World agriculture: Towards 2015/2030 – An FAO Perspective'', FAO, Earthscan Publications Ltd, London.

Cai, W. et al. (2015), "Increased frequency of extreme La Niña events under greenhouse warming", *Nature Climate Change*, Vol. 5, pp.132–137. www.nature.com/nclimate/journal/v5/n2/full/nclimate2492.html

California Employment Development Department (2015), "Statewide Historical Annual Average Employment by Industry Data, 1990–2014, Sacramento. Consulted July 2015. www.labormarketinfo.edd.ca.gov/LMID/Employment_by_Industry_Data.html

Carleton, T.A. and S.M. Hsiang (2016), "Social and economic impacts of climate", *Science,* Vol. 353/ 6304, aad9837. DOI: 10.1126/science.aad9837

CCTV (2016a), "El Nino brings floods in South China", CCTV, 16 June 2016, http://english.cctv.com/2016/05/16/VIDEZT3zLFMOSxqQbQ6zHOuL160516.shtml

CCTV (2016b), "Drought continues in central and northern China", 29 June 2016, http://english.cctv.com/2016/05/29/VIDETCDpbzjo4rDu54VD6FYv160529.shtml

CDFA (California Department of Food and Agriculture) (2015), "California Agricultural Exports, 2014-15", CDFA, Sacramento, CA. https://www.cdfa.ca.gov/statistics/PDFs/AgExports2014-2015.pdf

CDWR (California Department of Water Resources) (2014), "Scenarios of Future California Water Demand Through 2050 Growth and Climate Change", Sacramento, California. http://www.waterplan.water.ca.gov/docs/cwpu2013/Final/vol4/data_analytical_tools/05Scenarios_Future_California_Water_Demand.pdf

CDWR (2013). "California Water Plan". CDWR, Sacramento, California. http://www.waterplan.water.ca.gov/docs/cwpu2013/Final/06_Vol1_Ch05_Managing_an_UncertainFuture.pdf

Changnon, S. (1999), "Impacts of 1997–98 El Niño– Generated Weather in the United States", *Bulletin of the American Meteorological Society*, Vol. 80/9, pp 1819-1827. http://journals.ametsoc.org/doi/pdf/10.1175/1520-0477(1999)080%3C1819%3AIOENOG%3E2.0.CO%3B2

Chen, L. (2008), "Minister Chen's Speech at Second Sino-Swiss Workshop on Flood Control and Drought Relief", Ministry of Water Resources, People's Republic of China, November 10th, 2008. www.mwr.gov.cn/english/speechesandarticles/chenlei/200812/t20081205_59296.html

Chen, L. (2009), "Keynote Speech at the Special Session on Extraordinary Natural Disasters and Risk Management of Water Infrastructures of the 5th World Water Forum ", Ministry of Water Resources, People's Republic of China. www.mwr.gov.cn/english/speechesandarticles/chenlei/200904/t20090417_59297.html

Childs, N. and Kiawu, J. (2009), "Factors Behind the Rise in RCS-09D-01 May 2009 Global Rice Prices in 2008", United States Department of Agriculture, May 2009.

China Daily (2015), "Drought hits NE China", 22 July. www.chinadaily.com.cn/china/2015-07/22/content_21376023_2.htm

China Water Risk (2016), Website. http://chinawaterrisk.org/ (accessed February 2016).

Christian-Smith, J. et al. (2011), "Impacts of the California drought from 2007 to 2009", Pacific Institute, Oakland, CA. http://pacinst.org/app/uploads/2013/02/ca_drought_impacts_full_report3.pdf

City of Newport Beach (2014), "California disasters since 1950" in Natural Hazards Mitigation Plan, City of Newport Beach, California. www.newportbeachca.gov/Home/ShowDocument?id=19746

Climate Prediction Center (2016), "El Niño/Southern Oscillation (ENSO) diagnostic discussion", National Weather Service, NOAA. www.cpc.ncep.noaa.gov/products/analysis_monitoring/enso_advisory/ensodisc.pdf (accessed June 2016)

Cooley, H. et al. (2016), "Water Risk Hotspots for Agriculture: The case of the Southwest United States", OECD Food, Agriculture and Fisheries Paper N. 96, OECD Publishing, Paris.

Cooley, H. et al. (2015), "Impact of California's Ongoing Drought: Agriculture", Pacific Institute, Oakland, CA.

CRIDA (Central Research Institute for Dryland Agriculture) (2013), "Vision 2050", Central Research Institute for Dryland Agriculture, Santoshnagar, Hyderabad, India.

Darmouth Flood Observatory (2003a), "1988 Global Register of Extreme Flood Events", Dartmouth College, Hanover, NH. www.dartmouth.edu/~floods/Archives/1988sum.htm

Darmouth Flood Observatory (2003b), "2002 Global Register of Extreme Flood Events", Dartmouth College, Hanover, NH www.dartmouth.edu/~floods/Archives/2002sum.htm

Darmouth Flood Observatory (2006), "1998 Global Register of Extreme Flood Events", Dartmouth College, Hanover, NH www.dartmouth.edu/~floods/Archives/1998sum.htm

Dhaliwal, S. (2016), "Centre's drought aid to other states, no money for Punjab", The Tribune, India, 1st January. www.tribuneindia.com/news/punjab/centre-s-drought-aid-to-other-states-no-money-for-punjab/178016.html

Di Liberto, T. (2014), "ENSO and the Indian Monsoon… not as straightforward as you'd think", NOAA ENSO blog, www.climate.gov/news-features/blogs/enso/enso-and-indian-monsoon%E2%80%A6-not-straightforward-you%E2%80%99d-think

Dismukes, R. and K.H. Cobles (2006), "Managing risk with revenue insurance", *Amber Waves*, USDA Economic Research Services, Washington DC. www.ers.usda.gov/amber-waves/2006-november/managing-risk-with-revenue-insurance.aspx#.VvBdHSv0-yA

Döll, P. (2002), "Impact of climate change and variability on irrigation requirements: a global perspective", *Climatic change*, Vol. 54, pp. 269-293.

Downey, D. (2016), "Godzilla El Niño left Southern California mostly high and dry", *Los Angeles Daily News*, www.dailynews.com/environment-and-nature/20160609/godzilla-el-nino-left-southern-california-mostly-high-and-dry

ECA (Economics of Climate Adaptation Working Group) (2009), "Shaping Climate-Resilient Development: A Framework for Decision-Making", Report of the ECA Working Group, a partnership of Climate Works Foundation, Global Environment Facility, European Commission, McKinsey & Company, The Rockefeller Foundation, Standard Chartered Bank, and Swiss Re. http://ec.europa.eu/development/icenter/repository/ECA_Shaping_Climate_Resilent_Development.pdf

FAO (2015), The Impact of Disasters on Agriculture and Food Security, FAO, Rome. http://www.fao.org/resilience/resources/resources-detail/en/c/346258/

FAO (2012), "World agriculture towards 2030/2050: The 2012 Revision", Global Perspective Studies Team, ESA Working Paper No. 12-03, FAO, Rome.

FAO (1998), "Special Report: Floods cause extensive crop damage in several parts of Asia", Food and Agriculture Organization, Rome. www.fao.org/docrep/004/w9548e/w9548e00.htm#E62E3

Frisvold, G.B. et al. (2013), "Agriculture and Ranching." In Garfin, G., et al, Assessment of Climate Change in the Southwest United States: A Report Prepared for National Climate Assessment, edited by, 218–239, Southwest Climate Alliance, Island Press, Washington, DC. http://www.swcarr.arizona.edu/sites/default/files/ACCSWUS_Ch11.pdf

Frisvold, G.B. and K. Konyar (2012), Less Water: How Will Agriculture in Southern Mountain States Adapt? *Water Resources Research*, Vol. 48/5, W05534. doi:10.1029/2011WR011057.

Gleditsch N.P.(2012), "Special Issue: Climate Change and Conflict", *Journal of Peace Research*, Vol 49/1, pp. 3-9.

Gleick, P.H. (2014), "Water, Drought, climate change, and conflict in Syria", *Weather, Climate and Society*, Vol. 6/3, pp. 221-340.

Government of Haryana (2011), ''Haryana State Action Plan on Climate Change'', Government of Haryana, Chandigarh.

Government of Punjab (2016), "Draft Punjab State Disaster Management Plan", Chandigarh. http://punjabrevenue.nic.in/sdmp1234.pdf

Grogan, D.S. et al. (2015), "Quantifying the link between crop production and mined groundwater irrigation in China", *Science of the Total Environment*, Vol. 511, pp. 161-175.

G20 (2017), "G20 Agriculture Ministers' Declaration 2017-Towards food and water security: Fostering sustainability, advancing innovation", G20 German Presidency, Berlin, January 22 2017. http://www.bmel.de/SharedDocs/Downloads/EN/Agriculture/GlobalFoodSituation/G20_Declaration2017_EN.pdf;jsessionid=48AF37D05D7C0366B55297B7217593BE.2_cid376?__blob=publicationFile

Headey, D. and S. Fan (2011), "Reflections on the global food crisis: How did it happen? How has it hurt? And how can we prevent the next one?", IFPRI Research Monograph 165, International Food Policy Research Institute, Washington, DC.

Hejazi, M. I. et al. (2014), "Integrated assessment of global water scarcity over the 21st century under multiple climate change mitigation policies", *Hydrology and Earth System Sciences*, Vol. 18, pp.2859–2883. doi:10.5194/hess-18-2859-2014

Hidalgo, F.D. et al. (2010), "Economic Determinants of Land Invasions", *Review of Economics and Statistics*, Vol. 92, No. 3, pp. 505-523.

Hira, G.S. (2009), "Water Management in Northern States and the Food Security of India", *Journal of Crop Improvement,* 23/2, pp.136-157.

Hochman, Z. et al. (2017), "Climate trends account for stalled wheat yields in Australia since 1990", *Global Change Biology*, Vol. 23, pp. 2071-81. doi: 10.1111/gcb.13604

Howitt, R. et al. (2015), "Economic Analysis of the 2015 Drought for California Agriculture", UC Davis Center for Watershed Sciences, Davis, California. https://watershed.ucdavis.edu/files/biblio/Final_Drought%20Report_08182015_Full_Report_WithAppendices.pdf

Howitt, R.E. et al. (2014), "Economic analysis of the 2014 drought for California agriculture", UC Davis Center for Watershed Sciences, Davis, California. https://watershed.ucdavis.edu/files/content/news/Economic_Impact_of_the_2014_California_Water_Drought.pdf

Howitt, R. E., et al. (2009), "Estimating economic impacts of agricultural yield related changes", Final Report CEC-500-2009-042-F, California Climate Change Center, Sacramento, California.

Hundal S. and Kaur, P. (2006), "Effect of Possible Futuristic Climate Change Scenarios on Productivity of Some kharif and rabi Crops in the Central Agroclimatic Zone of Punjab", *Journal of Agricultural Physics*, Vol. 6/1, pp. 21-27.

ICAR (Indian Council for Agricultural Research) (2015), « Vision 2050 », Indian Agricultural Research Institute, New Delhi.

IRD (Institut de Recherche pour le Développement) (2013), "Le cousin indien d'El Niño : le 2e enfant terrible du climat", Institut de Recherche pour le Développement, Fiches d'actualités, Paris. www.ird.fr/la-mediatheque/fiches-d-actualite-scientifique/446-le-cousin-indien-d-el-nino-le-2e-enfant-terrible-du-climat

IRI (2016), "CPC/IRI Official Probabilistic ENSO Forecast", International Research Institute for Climate and Society, consulted version June 2016, http://iri.columbia.edu/our-expertise/climate/forecasts/enso/current/?enso_tab=enso-cpc_plume

Jägerskog, S. et. al. (2016), "Water, migration and how they are interlinked", Working paper No. 27, Stockholm International Water Institute (SIWI), Stockholm.

Jalota, S. et al. (2014), "Location specific climate change scenario and its impact on rice and wheat in Central Indian Punjab", *Agricultural Systems,* Vol. 131, pp. 77–86.

Johansson, E.L. et al. (2016), "Green and blue water demand from large-scale land acquisitions in Africa", *Proceedings of the National Academy of Sciences*, Vol. 13, N. 41, pp. 11471-11476. DOI:10.1073/pnas.1524741113

Kawashima, H. (2012), ''Influence of shortage of water resources on rice production in India: A district-level analysis'', Actualities of Indian Economic Growth at Rural-Urban Crossroads, University of Tokyo. http://www.l.u-tokyo.ac.jp/~tindas/fullpaper_kawashima.pdf

Keller, A.A. et al. (1996), "Integrated Water Resources Systems: Theory and Policy Implication", Research Report No 3, International Water Management Institute, Colombo, Sri Lanka.

Kelley, C.P. et al. (2015), "Climate change in the Fertile Crescent and implications of the recent Syrian drought", *Proceedings of the National Academy of Sciences*, Vol. 112/11, pp. 3241-3246

Krishan, G. et al. (2015), "Rainfall Trend Analysis of Punjab, India using Statistical Non-Parametric Test", *Current World Environment,* Vol. 10/3, pp. 792-800.

Krishnan, R et al. (2003), "Pacific decadal oscillation and variability of the Indian summer monsoon rainfall", *Climate Dynamics*, Vol. 21/3, pp 233–242. http://link.springer.com/article/10.1007/s00382-003-0330-8

Lau, K-M. and Weng H. (2001), "Coherent Modes of Global SST and Summer Rainfall over China: An Assessment of the Regional Impacts of the 1997–98 El Niño", *Journal of Climate*, Vol. 14/6. https://doi.org/10.1175/1520-0442(2001)014<1294:CMOGSA>2.0.CO;2

Lesk, C. et al. (2016), "Influence of extreme weather disasters on global crop production", *Nature*, Vol. 529, pp.84-87.

Levrault O. (2016), "Climat : Après El Niño, La Niña entrera en jeu à partir de juillet-août", Le Monde, 20 June. www.lemonde.fr/planete/article/2016/05/20/climat-apres-el-nino-la-nina-entrera-en-jeu-a-partir-de-juillet-aout_4923497_3244.html#LOosykhJTV8fVMET.99

Liu, J. et al. (2014), "International trade buffers the impact of future irrigation shortfalls", *Global Environmental Change*, Vol. 29, pp. 22–31. doi:10.1016/j.gloenvcha.2014.07.010

Lobell, D-B. et al. (2009), "Climate Extremes in California Agriculture", California Climate Change Center, Sacramento. www.energy.ca.gov/2009publications/CEC-500-2009-040/CEC-500-2009-040-F.PDF

Lobell, D.B. and C.B. Field (2011), "California perennial crops in a changing climate", *Climatic Change*, Vol. 109, Suppl. 1, pp. S317-S333.

Luedeling E. et al. (2009), "Climatic Changes Lead to Declining Winter Chill for Fruit and Nut Trees in California during 1950–2099", *PLoS ONE*, Vol. 4/7, e6166. doi:10.1371/journal.pone.0006166

MacDonald, G. M. (2010), "Water, climate change, and sustainability in the Southwest", *Proceedings of the National Academy of Sciences*, Vol. 107, pp. 21256-21262.

Marshall, E., et al. (2015), "Climate Change, Water Scarcity, and Adaptation in the U.S. Fieldcrop Sector", Economic Research Report No. (ERR-201), USDA Economic Research Service, Washington DC. http://ers.usda.gov/publications/err-economic-research-report/err201.aspx

Mauger, G. et al. (2015), "Impacts of Climate Change on Milk Production in the United States", *The Professional Geographer,* Vol. 67/1, pp. 121–31. doi:10.1080/00330124.2014.921017.

Michael, J. et al. (2010), " A Retrospective Estimate of the Economic Impacts of Reduced Water Supplies to the San Joaquin Valley in 2009", University of the Pacific, California. www.pacific.edu/Documents/school-business/BFC/SJV_Rev_Jobs_2009_092810.pdf

Ministry of Earth Sciences (2016), "El Nino/La Nina Indian Ocean Dipole Update (10th June 2016)", Government of India, New Delhi. www.imdpune.gov.in/Clim_RCC_LRF/ENSO_Bulletin/ENSO_IOD_Update_Bulletin_Jun16.pdf

Mohanti, S. (2016), ''Impact of El Niño? The rice market is not bothered'', Sam's rice price and market blog, International Rice Research Institute, Los Baños, Philippines. http://irri.org/blogs/sam-s-rice-price-and-market-blog/impact-from-el-nino-the-rice-market-is-not-bothered

Mukherjee, S. (2015), "Excess rain in March a departure from trend", *Business Standard*, 4 April. www.business-standard.com/article/economy-policy/excess-rain-in-march-a-departure-from-trend-115040300725_1.html

Naumann, G. et al. (2015), "Assessment of drought damages and their uncertainties in Europe", *Environmental Research Letters*, Vol. 10/12. http://iopscience.iop.org/article/10.1088/1748-9326/10/12/124013/pdf

New York Times (1998), "El Nino Brings Flooding and High Winds to Coastal California", 4 February. www.nytimes.com/1998/02/04/us/el-nino-brings-flooding-and-high-winds-to-coastal-california.html

Nillesen, E. and E. Bulte (2014), "Natural Resources and Violent Conflict", *Annual Review of Resource Economics*, Vol. 6, pp. 69-83.

NPR (National Public Radio) (2015), "Saudi Hay Farm In Arizona Tests State's Supply Of Groundwater", NPR The Salt, Washington DC. http://www.npr.org/sections/thesalt/2015/11/02/453885642/saudi-hay-farm-in-arizona-tests-states-supply-of-groundwater

Null J. (2004) "An Analysis of El Niño, La Niña and California Rainfall", Golden Gate Weather Services, Saratoga, California. http://ggweather.com/enso/calenso.htm

Null J. (2015), "The Misconceptions of El Niño", Golden Gate Weather Services, Saratoga, California. http://ggweather.com/enso/enso_myths.htm

OECD (2016), *Mitigating Droughts and Floods in Agriculture: Policy Lessons and Approaches*, OECD Studies on Water, OECD Publishing, Paris, http://dx.doi.org/10.1787/9789264246744-en.

OECD (2015), *The Economic Consequences of Climate Change*, OECD Publishing, Paris, http://dx.doi.org/10.1787/9789264235410-en .

OECD (1999), "Grain Production and Income of Farmers in China", in Agriculture in China and OECD Countries, https://one.oecd.org/document/2348/en/pdf

OECD/FAO (2015), Figure 1.15., ''Concentration of exports by commodity, 2024'', OECD-FAO Agricultural Outlook 2015, OECD Publishing, Paris, http://dx.doi.org/10.1787/agr_outlook-2015-en.

Office of the Director of National Intelligence (2012), "Global Water Security", Intelligence Community Assessment, National Intelligence Council, Washington DC.

Paulson, R.W. et al. (1990), "National Water Summary 1988-89-- Hydrologic Events and Floods and Droughts", U.S. Geological Survey Water-Supply Paper 2375, USGS, Washington DC. http://geochange.er.usgs.gov/sw/impacts/hydrology/state_fd/cawater1.html

Pearce, F. (2013), "Splash and grab: the global scramble for water", *The New Scientist*, 27 February. https://www.newscientist.com/article/mg21729066-400-splash-and-grab-the-global-scramble-for-water

Piao, S. et al. (2010), "The impacts of climate change on water resources and agriculture in China", *Nature*, Vol. 467, pp.43-51.

Post, R. et al. (2016), "Rethinking the water-food-climate nexus and conflict: An opportunity cost approach", *Applied Economic Perspectives and Policy*, Vol. 38, pp.563-577.

Rao, S. (2015), "Indian Ocean Dipole and the monsoon: The Joker in the forecast pack" (interview), written by Nambiar, N., *The Indian Express*, 8 June 2015, http://indianexpress.com/article/explained/indian-ocean-dipole-and-the-monsoon-the-joker-in-the-forecast-pack/

Robinson, S. et al. (2015) "The International Model for Policy Analysis of Agricultural Commodities and Trade (IMPACT); Model description for version 3", IFPRI Discussion Paper, International Food Policy Research Institute, Washington, DC.

Rosenzweig, C., et al. (2004), "Water resources for agriculture in a changing climate: International case studies", *Global Environmental Change A*, Vol. 14, pp. 345-360. doi:10.1016/j.gloenvcha.2004.09.003.

Rulli, M.C. et al. (2013), "Global land and water grabbing", *Proceedings of the National Academy of Sciences*, Vol. 110/3, pp. 892-897.

Sadoff, C. et al. (2015), "Securing Water, Sustaining Growth", Report of the GWP/OECD Task Force on Water Security and Sustainable Growth, Oxford University, Oxford. https://www.water.ox.ac.uk/wp-content/uploads/2015/04/SCHOOL-OF-GEOGRAPHY-SECURING-WATER-SUSTAINING-GROWTH-DOWNLOADABLE.pdf

Seckler, D. et al. (1999), "Water Scarcity in the Twenty-first Century", *International Journal of Water Resources Development*, Vol. 15/1-2, pp. 29-42.

Sekhri, S. (2014), "Wells, Water, and Welfare: The Impact of Access to Groundwater on Rural Poverty and Conflict", *American Economic Journal: Applied Economics*,Vol. 6/3, pp.76–102.

Shiao, T. et al. (2015), "India Water Tool", technical note, World Resources Institute, February 2015. http://www.wri.org/resources/maps/india-water-tool

Shiklomanov, I. A. (1999) Electronic data provided to the Scenario Development Panel, World Commission on Water for the 21st Century, World Water Council, Marseille.

Shiva, V. (1991), *The Violence of the Green Revolution*, University Press of Kentucky, 144 p.

Shrivastava, A. (2016), ''Climate change and Indian agriculture'', *International Policy Digest*, 22 August. http://intpolicydigest.org/2016/08/22/climate-change-and-indian-agriculture/

Siegel, S. (2015), *Let there be water: Israel's solution for a water-starved world*, St Martin's Press, New York.

Skinner, J. and L. Cotula (2011), "Are land deals driving 'water grabs'?", IIED Briefing, November, International Institute for Environment and Development, New York. http://pubs.iied.org/pdfs/17102IIED.pdf

Smaller, C. and Mann, H. (2009), "A Thirst for Distant Lands: Foreign investment in agricultural land and water", International Institute for Sustainable Development, Winnipeg, Canada.

Soora, N.K. et al. (2013), "An assessment of regional vulnerability of rice to climate change in India", *Climatic Change*, Vol.118/3, pp. 683–699.

Spagat, E., and A. Batrawy (2016), "Saudi Arabia buys California farmland... for the water", the Associated Press, 28/03/2016. http://www.dailynews.com/article/20160328/NEWS/160329565

Srinhidi, A. et al. (2015), "Lived anomaly - How to enable farmers in India to cope with extreme weather events", Centre for Science and Environment, New Delhi. http://cdn.downtoearth.org.in/pdf/lived-anomaly-extreme-weather-events.pdf

State of California (2012), "Flood Damage Analysis" in 2012 Central Valley Flood Protection Plan (Public Draft), State of California, Department of Water resources, Sacramento, California. www.water.ca.gov/cvfmp/docs/Att8F_FloodDamage_20120224.pdf

Tao, F. et al. (2003), "Future climate change, the agricultural water cycle, and agricultural production in China", *Agriculture, Ecosystems and Environment*, Vol. 95, pp. 203–215.

The Times of India (2013), "Government defers promulgation of ordinance on Food Security Bill", 13 June. http://timesofindia.indiatimes.com/india/Govt-defers-promulgation-of-ordinance-on-Food-Security-Bill/articleshow/20569710.cms

Totten, S. (2015), "Forget El Niño, the 'PDO' could be the real drought buster", Southern California Public Radio, 13 August. www.scpr.org/news/2015/08/12/53755/forget-el-nino-could-the-pdo-be-the-real-drought-b/

United Nations (2010), "Foreign land purchases for agriculture: what impact on sustainable development?", Sustainable Development Innovation Briefs, UN, New York.

UNISDR and CRED (UN Office for Disaster Risk Reduction and Center for Research on the Epidemiology of Disasters) (2015), "The Human Cost of Weather Related Disasters 1995-2015", UNISDR, Geneva and CRED, Louvain, Belgium. https://www.unisdr.org/we/inform/publications/46796

USACE (U.S Army Corps of Engineers) (1993), "Lessons Learned from the California Drought (1987-1992): National Study of Water Management During Drought", Institute for Water Resources, Report 93-NDS-5, USACE, Washington DC. http://www.iwr.usace.army.mil/Portals/70/docs/iwrreports/93-NDS-5.pdf

USBR (2012), "Colorado River Basin Water Supply and Demand Study", USBR, Washington DC. http://www.usbr.gov/lc/region/programs/crbstudy/finalreport/Study%20Report/CRBS_Study_Report_FINAL.pdf

Von Braun, J. and R.S. Meinzen-Dick (2009), "'Land grabbing' by foreign investors in developing countries: risks and opportunities", IFPRI Policy Brief 9, International Food Policy Research Institute, Washington DC. http://ebrary.ifpri.org/utils/getfile/collection/p15738coll2/id/14853/filename/14854.pdf

Wada, Y. et al. (2013), "Multimodel projections and uncertainties of irrigation water demand under climate change", *Geophysical Research Letters*, Vol. 40, pp. 4626–4632. doi:10.1002/grl.50686

Walia, A. and Sharma, J. (2015), ''Free power, the bane of farming in Punjab'', *The Hindu Buisness Line*, 16 August. www.thehindubusinessline.com/opinion/free-power-the-bane-of-farming-in-punjab/article7546918.ece

Wang, J., et al. (2014), "Overview of Impacts of Climate Change and Adaptation in China's Agriculture", *Journal of Integrative Agriculture*, Vol. 13/1, pp. 1-17.

WEF (World Economic Forum) (2017), "Global Risks 2017", 12th report, WEF, Geneva.

WEF (2016), "Global Risks 2016", 11th report, WEF, Geneva.

WEF (2015), "Global Risks 2015", 10th report, WEF, Geneva.

Weiss, K.R. (2015), "The making of a climate refugee", *Foreign Policy*, 28 January. http://foreignpolicy.com/2015/01/28/the-making-of-a-climate-refugee-kiribati-tarawa-teitiota/

Werz, M. and L. Conley (2012), "Climate change, migration and conflict", Center for American Progress, Washington DC.

Willenbockel, D. (2012), "Extreme weather events and crop price spikes in a changing climate", Oxfam Research Reports, Oxfam International, Oxford, UK. https://www.oxfam.org/sites/www.oxfam.org/files/rr-extreme-weather-events-crop-price-spikes-05092012-en.pdf

Woodhouse, P. (2012), "New investment, old challenges. Land deals and the water constraint in African agriculture", *The Journal of Peasant Studies*, Vol. 39/3-4, pp. 777-794.

World Bank (2014), "Republic of India - Accelerating Agricultural Productivity Growth", World Bank Group Agriculture, Washington DC. http://www-wds.worldbank.org/external/default/WDSContentServer/WDSP/IB/2015/01/14/000442464_20150114081222/Rendered/PDF/880930REVISED00ivity0Growth00PUBLIC.pdf

World Bank (2012), "Severe droughts drive food prices higher, threatening the poor", Press Release, August 30, 2012, World Bank, Washington, D.C. http://www.worldbank.org/en/news/press-release/2012/08/30/severe-droughts-drive-food-prices-higher-threatening-poor

Xiong, W., et al. (2010), "Climate change, water availability and future cereal production in China", *Agriculture, Ecosystems and Environment*, Vol. 135, pp. 58-69.

Xiong, W, et al. (2009), "Future cereal production in China: The interaction of climate change, water availability and socio-economic scenarios", *Global Environmental Change*, Vol. 19, pp. 34-44.

Xurong, M. and Y. Xiaohui (2012), "Drought conditions and management strategies in China", UN Water, New York. www.ais.unwater.org/ais/pluginfile.php/597/mod_page/content/79/China.pdf

Yao, Y. (2007), "Studies on China's social-economic scenarios in 2005-2050", Annual report of Chinese Academy of Social Science, Beijing.

Yu, M. et al. (2014), "Are droughts becoming more frequent or severe in China based on the Standardized Precipitation Evapotranspiration Index: 1951–2010?", Drought Mitigation Center Faculty Publications, University of Nebraska – Lincoln, Lincoln, Nebrasaka. http://digitalcommons.unl.edu/cgi/viewcontent.cgi?article=1016&context=droughtfacpub

Yuan, Y. and Yang, S. (2012), "Impacts of Different Types of El Niño on the East Asian Climate: Focus on ENSO Cycles", *Journal of Climate*, Volume 25/21. https://doi.org/10.1175/JCLI-D-11-00576.1

Zetland, D. and Möller-Gulland, J. (2012), "The political economy of land and water grabs", in Allan, T. et al. (eds), *Handbook of Land and Water Grabs*, Routledge.

Annex 3.A1

Modelling scenarios and assumptions

The model used for this study is the multi-market partial equilibrium model IMPACT 3.1, which enables projecting both climate change and water stress at a subnational level on specific commodities (Robinson et al., 2015). The projection of irrigation water demand depends on changes in irrigated area and cropping patterns, water use efficiency, and rainfall harvest technology. Global climate change can also affect future irrigation water demand through changes in temperature and precipitation.

More precisely, irrigation water demand and supply are not formally defined in this model, as they are exogenous variables coming from the Impact Water Model Simulation Model (IWSM). The variation in water availability for agriculture year by year in different climate scenarios has an impact on yields. This mechanism is modelled through the use of the IMPACT water models. These include (1) a global hydrology model that determines runoff to the river basins included in the IMPACT model; (2) water basin management models for each food production unit (FPU, the unit of modelling, combining watersheds with agriculture production areas) that optimally allocate available water to competing non-agricultural and agricultural uses, including irrigation; and (3) a water allocation and stress model that allocates available irrigation water to crops and, when the water supply is less than demand by crop, computes the impact of the water shortage on crop yields, accounting for differences among crops and varieties. These yield shocks are then passed to the IMPACT model, affecting year-to-year crop yields (Robinson et al., 2015).

The following sections summarise the scenarios and assumptions of the proposed simulations. A subsequent section then reviews the evidence of impact and occurrence of extreme precipitation patterns in the three regions.

3.A1.1. Geographical and agriculture scope

The three regions are defined as follow in the simulations:

- Northeast China covers four main river basins. The Huang He, also known as the Yellow River watershed, contains the country's second largest river. The Huang He rises in the southern Qinghai province on the Plateau of Tibet. Another river basin, the Huai He is located about halfway between the Yellow River and Yangtze, the latter being the two largest rivers in China. The Huai He is considered the geographical dividing line between Northern and Southern China. The last two watersheds modelled are the Hai He basin and the Songhua basin, which are drained by the Liao River. Maize, vegetables, wheat, potatoes and cotton are the main crops produced in Northeast China.

- Three major water river basins can be delimited in Northwest India: the Indus, the Luni, and the Ganges. The Indus river basin spreads over states of Jammu & Kashmir, Himachal Pradesh, Punjab, Rajasthan, Haryana and Union Territory of Chandigarh with an area of 321 289 km^2, nearly 9.8% of the total geographical area of India. The Luni watershed includes the western part of Rajasthan and the northern part of Gujarat. The Ganges River is the third river basin concerned, covering a very large share of Northern India, but is left aside from the simulation as much of this region lies outside of Northwest India. Sugar cane, wheat, rice, potatoes, subtropical fruits are the main crops produced in Northwest India.

- Three main watersheds delimit the Southwest United States: the California River Basin, the Colorado River Basin, and the Great Basin. The Colorado River Basin, houses the largest river in the Southwest, and includes parts of California, Utah, Arizona, New Mexico, Colorado, and Wyoming. The Great Basin watershed includes parts of Southern California, most of Utah, the northwest half of Nevada, and parts of both Oregon and Idaho. Dairy, cattle (beef), maize, fruits, vegetables, wheat and potatoes are the main commodities produced in United States Southwest.

3.A1.2. Modelling water stress on irrigation in IMPACT

The baseline scenario is defined by assuming no climate change (no RCP) and no adaptation policies. This scenario is referred to as BAU (Business as usual) and is noted S0. For water stress simulations, we implement three scenario variations–basin efficiency deficit (SB) and groundwater depletion (SG) and drought (SD). Two regional climate scenarios (RCP 8.5 IPSL and RCP 8.5 Hadley) are added. This equates to a possible twelve scenario runs (three climate options * four combinations of water stress).

The two reference climate scenarios, which are defined by no water stress simulations, are referred to as Hadley S1 and IPSL S1. Thus, the specific effects of the two water stress simulations on agriculture are readily analysed by comparing each additional water stress scenario (SD, SB, SG) to the reference scenario (S0). Table 3.A1.1 describes the scenarios implemented in the study.

Table 3. A1.1. Specification of the key water stress scenarios

Scenario	Measure specification	Regional scope / river basins	Timing	Activities affected
Deficit in basin efficiency Scenario (SB)	50% decreased basin water use efficiency by 2050 compared with 2005 levels, until all regional basins reach a maximum efficiency of 46%.	*Southwest United States* California basin; Colorado river basin; Great basin	2005-50	All
	50% decreased basin water use efficiency by 2050 compared with 2005 levels, until all regional basins reach a maximum efficiency of 33%.	*Northwest India* Indus basin; Luni basin	2005-50	All
	50% decreased basin water use efficiency by 2050 compared with 2005 levels, until all regional basins reach a maximum efficiency of 40%	*Northeast China* Hail He basin; Huai He basin; Huang He basin Songhua basin	2005-50	All
Groundwater depletion scenario (SG)	50% increased groundwater depletion by 2050 compared with 2005 levels.	*Southwest United States* California basin; Colorado river basin; Great basin	2005-50	All
	50% increased groundwater depletion by 2050 compared with 2005 levels.	*Northwest India* Indus basin; Luni basin	2005-50	All
	50% increased groundwater depletion by 2050 compared with 2005 levels.	*Northeast China* Hail He basin; Huai He basin; Huang He basin Songhua basin	2005-50	All
Drought scenario (SD)	Simulation of a drought that decreases the average yields of field crops by 30%, cattle and dairy by 20% vegetable by 15%, and fruits by 10%	*Southwest United States* California basin; Colorado river basin Great Basin	2021	Beef, dairy, maize, fruits and vegetables, wheat, potatoes
	Simulation of a drought that decreases field crop average yields by 30% and subtropical fruit average yield by 10%.	*Northwest India* Indus basin; Luni basin	2030	Sugar cane, wheat, rice, potatoes, subtropical fruits
	Simulation of a drought that decreases decrease average yields of field crop by 30%, vegetables by 15%	*Northeast China* Hail He basin; Huai He Basin; Huang He basin Songhua basin	2030	Maize, vegetable, wheat, rice, potatoes, cotton.

Source: Authors' own assumptions.

Technical specification of the basin deficiency scenario (SB)

The scenario on decreased basin efficiency is designed to mimic a deterioration of future water supplies usable for irrigation. Based on various definitions provided by Keller et al. (1996), the IMPACT model defines basin efficiency as the ratio of beneficial water depletion (crop evapotranspiration and salt leaching) to total irrigation water depletion at the basin scale. Basin efficiency in the base year (2005) is calculated as the ratio of the net irrigation water demand to the total irrigation water depletion estimated from records (Shiklomanov, 1999). The projection of irrigation water demand depends on the changes in irrigated area and cropping patterns, basin efficiency, and effective rainfall. Global climate change affects future irrigation water demand through changes in precipitation and temperature along with other meteorological variables that affect crop evapotranspiration. Irrigation demand in the FPU is calculated for a given cropping pattern after taking into account the basin efficiency of the irrigation system.

As simulated, reduced basin efficiency has a negative impact on surface water availability for irrigation. The assumed parameters do not aim to provide accurate or realistic projections, especially in the United States Southwest, but they are meant to implement water supply constraint shocks that could affect irrigated agriculture in the three regions. In other words, the scenarios provide a means to gauge the resistance of the irrigation sector to water stress (providing a "water stress test" like the other scenarios).

As shown in Table 3.A1.1, surface water constraint is simulated by assuming a 50% decrease in the baseline basin efficiency by 2050. This translates into simulating the gradual transition to a maximum of 46%, 40% and 33%, basin efficiency by 2050 in the Southwest United States, Northeast China and Northwest India, respectively. The shocks are intended to provide a pessimistic illustration of the possible impacts of agriculture water risks in the three regions, rather than an accurate projection of observed trends. This efficiency loss by 2050 is assumed to follow a linear decrease over the 45-year projection period, representing slow but steady decreases in basin efficiency. The basin efficiency trend for each FPU is assumed to be linear over the period and can be calculated as follows: $\frac{ee_{fpu,yrs}-ee_{fpu,2005}}{n}$ with $ee_{fpu,2005} > ee_{fpu,yrs}$ s with ee = Effective Basin Efficiency defined on *fpu* and *yrs* and n = number of periods (n=2050-2005=45).

There is no direct relationship between the yield parameter and basin efficiency (ee). Basin efficiency is used in the water allocation part of the IMPACT water models. Decreasing efficiency, all things equal, will decrease water supply in the basin by affecting water reliability and by increasing the risks associated with water scarcity. When water is a limiting factor, the impacts of basin efficiency variations are spread across all crops, and as such it is usually difficult to see major yield changes unless the change in efficiency is large or the water stress is severe. As the water allocation is optimised at the full basin (not the FPU) it is possible to see limited effects in yields in a basin where the efficiency decreases, as it may limit water to be used in other FPUs within the basin.

Technical specification of the groundwater depletion scenario (SG)

The second irrigation policy to be explored is groundwater overuse, where withdrawal of water is assumed to exceed recharge, leading to underground aquifers depletion. A 50% increase in groundwater depletion by 2050 compared with 2005 levels is modelled in all water basins. Similar to the irrigation efficiency trend, groundwater depletion is assumed to increase linearly over the 45-year projection period. It should be noted that both decreased basin efficiency and increased groundwater depletion could lead to lower yields, but will mostly affect water reliability and increase the risks associated with water scarcity.

Groundwater depletion is defined as the case when groundwater abstraction is greater than groundwater recharge, including return flows. For the desired 2050 groundwater reduction level to be reached, a groundwater withdrawal trend has to be estimated. The groundwater trend for each FPU is assumed to be linear over the period and can be calculated as follows: $\frac{gwd_{fpu,yrs}-gwd_{fpu,2005}}{n}$ $gwd_{fpu,2005}$ > Max $gwd_{fpu,yrs}$

with *gwd* = groundwater depletion defined on *fpu* and *yrs* and *n* is the number of periods (n=2050-2005=45).

Similar to basin efficiency, there is no direct relationship between yield and groundwater depletion (gwd). Groundwater is used in the water allocation part of the IMPACT water models. Increasing groundwater depletion, all other things equal, will decrease water supply in the basin. As the water allocation is optimised at the full basin (not the FPU) it is possible to see limited effects in yields in a basin where the groundwater depletion is higher, as it may limit water to be used in other FPUs within the basin. In practice, groundwater depletion may not impact groundwater withdrawal; in the Southwest United States, producers tend to update pumps to maintain their groundwater pumping (but in such case, it would affect their production costs).

Technical specification of the drought scenario (SD)

To evaluate the impact of water stress on irrigation in the three regions, a drought scenario can be elaborated. In that effect, an exogenous shock is applied on crop yields that will decrease to reflect the consequences of a drought, as seen in Table 3.A1.1. In IMPACT model, yields are estimated from initial yields, exogenous assumptions on yield growth (from technology, water, and climate shocks working through DSSAT and the water models), and current net price of the outputs, which takes into account the cost of inputs. The first three effects on yield are essentially exogenous to IMPACT, while the last one is endogenous and part of the IMPACT model equations.

$$Yield_{j,fpu,lnd} = YieldInt_{j,fpu,lnd} \times YieldInt2_{j,fpu,lnd} \times WatShk_{j,fpu,lnd} \times CliShk_{j,fpu,lnd} \times YieldShk_{j,fpu,lnd} \times \left(\frac{PNET_{j,cty}}{PNET0_{j,cty}}\right)^{Y\varepsilon}$$

With the following parameters:

Yield	Final Crop Yield
YieldInt	Crop yield intercept (basc year crop yield)
YieldInt2	Exogenous Crop yield growth multiplier due to technology
YieldShk	Exogenous yield shock
WatShk	Water stress shock (from water models) – continuous parameter
CliShk	Climate change shock (from water and crop models) – continuous parameter
PNET	Current Net Price
PNET0	Base Year Net Price (used to index prices in equation)
Yε	Yield supply elasticity with respect to net price

To model a drought in all basins, a yield decrease is simulated for various commodities by shocking the parameter *yieldshk*. In United States Southwest, yields decrease by 30% for crops (maize, wheat and potatoes), 20% for cattle and dairy, 15% for vegetables and 10% for fruits for the year 2021. In Northeast China, drought simulations decrease average yields of field crop (maize, wheat, rice, potatoes, cotton) by 30% and vegetables by 15% in 2030. In Northwest India, a 30% crop yield (sugar cane, wheat, rice, potatoes) decrease, and a 10% subtropical fruit yield decrease are simulated in 2030 to reflect a drought. These figures are based on yield decreases observed in the IMPACT model simulations under various climate change assumptions (Ignaciuk and Mason-D'Croz, 2014).

3.A1.3. Extreme precipitation patterns and agriculture production in the three regions

This section reviews past evidence on the damages of extreme precipitation patterns in the three regions, and provides a rapid assessment of possible future events.

Frequency of extreme precipitations in relation with El Niño Southern Oscillation (ENSO) : Pacific Decadal Oscillation (PDO) and Indian Ocean Dipole (IOD) in the three countries

In Northeast China, the drought frequency increased to 6.88% during 1977-2010 from the 2.80% observed during 1951-76 (Yu et al., 2014). The relationship between rainfalls and ENSO events are more complex. The influence in the North of the country is not clear. A study estimates that 30% of the annual rainfalls anomalies are related to ENSO in China, but more clearly in the South (Lau and Weng, 2001). It might depend of different types of *El Niño* (Central Pacific vs East Pacific) (Yuan and Yang, 2012). However, *El Niño* is associated with drought and La Nina with above normal precipitations in North China (Yu et al., 2014). At the same time, a negative Pacific Decadal Oscillation (PDO) strengthens the monsoon in the North (Browning, 2013).

In Northwest India, relatively strong trends have been associated with ENSO events but there have been exceptions. From 1950-2012, 80% of La Nina events (ENSO-) were associated with above normal precipitations during the monsoon (*kharif*) (Di Liberto, 2014). 71% of El Nino events (ENSO+) were associated with below normal precipitations during the monsoon (Di Liberto, 2014). El Nino is also associated with abnormal heavy rains and hails from February to April in Northwest India (Mukherjee, 2015), which is normally the dry season (*Rabi*). However, there have been significant anomalies: in 1997-98, the strongest *El Niño* of the century was associated with above normal precipitations, and one of the driest monsoons recorded, in 2002, was associated with a weak *El Niño*. At the same time a positive PDO is associated with drought and a negative PDO with above normal rainfalls when negative (Krishnan et al., 2003). A negative IOD (Indian Ocean Dipole) index is correlated with dry monsoon (Rao, 2015).

In the Southwest United States, and specifically California, strong ENSO+ events seem to be statistically associated with increased precipitations. The influence is more significant in Southern California. Trends in ENSO indices from 1950-2010 show that there was an even distribution in sub/normal/over precipitations in the North during the 23 *El Niño* events. But three of the five "strong" indices came with above normal precipitations (+120%)–the fifth below normal. In the central part of California, there was also an even distribution in sub/normal/over precipitations during the 23 *El Niño* events; but four of the five "strong" came with more than 140% above normal precipitations–the fifth below normal. In the South, there was also an even distribution in sub/normal/over precipitations during the 23 *El Niño* events; but four of the five "strong" came with more than 140% above normal precipitations–the fifth normal. (Null, 2015).[1] ENSO- is statistically associated with decreased precipitations, at least in Southern California. Observing the La Nina events from 1950-2000 in California, the following observations were made. In the North, over the 12 *La Niñas*, four were below normal (-20% and more), six normal, two above (+20% and more) (Ref Station San Francisco); in the South: Over the 12 *La Niñas*, seven were below normal (-20% and more, including 4 of -40% and more), five normal, zero above (+20% and more) (Null, 2004). PDO positive events were associated with drought, and PDO negative events with above normal rainfalls (Browning, 2013; Totten, 2015).

Impact of recent extreme events on agriculture in the three regions

A review of recent extreme weather events in the three regions provides a benchmark to the possible impact of future extreme events. Tables 3.A1.2 to 3.A1.6 offer a rapid characterisation of the main climatic phenomena in the three regions since 1988 and their reported impact on agriculture.

Table 3.A1.2. Impacts of climatic events in the three hotspot regions: 1988-1989.

1988-1989: Strong negative Oceanic *Niño* Index –PDO hot- IOD 0		
India (NW)	**United States (California)**	**China (NE)**
- Very severe floods in September 1988: 1 463 villages inundated in Punjab, 381 in Haryana (Dartmouth Flood Observatory, 2003a). - 80% of standing crops destroyed in Punjab (Shiva, 1991).	- Drought t begun in 1987 and lasted over 1988 and 89. Extremely severe US-wide. - In California, 15 to 40 recurrence interval depending on areas (Paulson et al., 1990). - USD 500M loss in farm sales (USACE, 1993).	- Persistent rains attributed to *La Nina;* USD 150 million farm loss country-wide (FAO, 1998). - Including, in August 1998 : Eastern Shandong Province, Yellow River flooded 33 villages (16810 ha); Shanxi (36880 ha flooded); Henan (153100 ha) (Dartmouth Flood Observatory, 2006).

Source: Authors' own work based on cited references.

Table 3.A1.3. Impacts of climatic events in the three hotspot regions: 1997-1998

1997-1998: Very strong Oceanic Niño Index- PDO hot – Very high IOD		
India (NW)	**United States (California)**	**China (NE)**
-1998: India's deadliest heat wave. The monsoon precipitations were normal. This could be explained by the high IOD (Indian Ocean Dipole) index, correlated with a good monsoon. There was no significant drop in agriculture production (World Bank, 2014). -The year is an example of ambiguity: even if ENSO+ is statistically associated with droughts in India, this is not systematic. (Levrault, 2016).	-Floods in California in February 1998, associated with El Nino (New York Times, 1998). The exceedance probability of the February 23 flood peak reached 4%, equivalent to a one in 25-year flood (Bowers, 2001). -February 1998 brought several record-breaking rainfalls (Greatest rainfall since 1877), with 50-year storm event intensities (Dartmouth Flood Observatory, 2006). -March: Flood in Monterey, San Luis Obispo, San Benito, Napa: estimated 342M$ crop loss; May: Heavy rain in central California, estimated USD 310 M crop loss; June: flood in Southern San Joaquin Valley, estimated crop loss USD 100M (Lobell et al., 2009). -The floods in January 1997, not under El Nino conditions, were worse (1.8B$ total costs vs 0.55B$) (City of Newport Beach, 2014). -There is a large difference in Californian losses between similar El Niño; USD 2 billion in 1982–83 versus USD 1.1 billion in 1997–98 (Changnon, 1999).	-Beginning of a five years drought in NE China. -1997: catastrophic drought in the Yellow river basin (226 days zero flow) (Yu et al., 2014); up to 60 million metric tons grain loss. -2.5% yield loss (grain) because of drought (OECD, 1999).

Source: Authors' own work based on cited references.

Table 3.A1.4. Impact of climatic events in the three hotspot regions: 2002-2003

2002-2003 : Moderately strong Oceanic Niño Index – PDO cold-IOD 0		
India (NW)	**United States (California)**	**China (NE)**
Very severe drought: 22% aggregate rainfall anomaly and -25% trend deviation in agriculture production (World Bank, 2014). -All Punjabi districts affected, reported as the most severe (Government of Punjab, 2016).	Storms and heavy rains end of December 2002, partly due to Nino conditions in the Pacific. Affected 65 000 km2, including the long-term inundation of San Joaquin and Sacramento Valley (Dartmouth Flood Observatory, 2003b; State of California, 2012).	-No particular drought. -Amelioration of the hydric conditions after several years of drought in Northern China (500-600. 10*5 tonnes crop loss /year in 2000 and 2001 due to drought (Xurong and Xiaohui, 2012). -Could be explained by shift to cold PDO, which increases monsoon in the north and decreases it in the South (Browning, 2013).

Source: Authors' own work based on cited references.

Table 3.A1.5. Impact of climatic events in the three hotspot regions: 2007-2008

2007-2008 : Strong Negative Oceanic Niño Index (La Nina) – PDO cold- IOD 0		
India (NW)	**United States (California)**	**China (NE)**
-Severe floods, including in the NW, in August 2007 and 2008. Floods in Punjab 2007: INR 582995000 (= USD 8.6M) crop damages. 2008: INR 645084000 (=9.6M$) (2nd and 3rd most expensive since at least 1960. The worst flood occurred in 1995) (Government of Punjab, 2016).	- Drought in California : USD 20 M USDA drought-related crop insurance payments in California during the 2007-2009 drought (Christian-Smith et al., 2011). - While 2007 and 2008 were dry years, they are among the decade's highest in terms of yield of irrigated crops state-wide. - Still, USD 368 M was lost due to drought and pumping restrictions across the entire San Joaquin Valley. This represents a 2.5% decline in revenue across the Valley. USD 328 M (89%) is in Kern and the west-side regions. (Michael et al., 2010).	-Severe floods from June to August 2007. The 2007 flood volume of Huai River is the second largest since1949, next to that of 1954 basin. It represented three times the normal average (Chen, 2008). -In the summer 2007, there was a severe drought in Southern China (Dartmouth Flood Observatory, 2003) (inversely, El Nino bring floods in South, like in 1997-8 and 2015-16) (CCTV, 2016a). -In 2008, 10 typhoons or tropical storms hit China (vs 7 in average) (Chen, 2009).

Source: Authors' own work based on cited references.

Table 3.A1.6. Impact of climatic events in the three hotspot regions: 2015-16

2015-2016: Very strong Oceanic Niño Index – PDO hot- IOD+		
India (NW)	**United States (California)**	**China (NE)**
-Heat records in 2015 – 2016 -dry kharif season in 2015: 40% rain deficit in Punjab, 38% in Haryana (Dhaliwal, 2016). -Unseasonal rains and hails during rabi 2015 (biggest losses: Haryana and Rajasthan, which led to a 7% drop in wheat production in India 2015 (to 86.53mt) (Srinhidi, et al., 2015); the standing crop worth Rs717 crore was damaged over an area of 294,177 hectares in Punjab) (Bhatnagar, 2015). The same phenomenon occurred in 2016.	-Expected increase in rainfalls because *El Niño* after four years of drought (Howitt et al., 2015) did not occur (Downey, 2016). -USD 900M$ crop loss, USD 350M livestock loss, USD 590M in additional pumping in California (Howitt et al., 2015).	-Droughts in central and Northern China 2015-2016 (CCTV, 2016b). -Drought summer 2015 affected 1.15 million hectares of crops in NE (China Daily, 2015). -Drought in the summer 2016 affected over 1 million hectare of crops (Hebei, Henan and Shaanxi provinces) (CCTV, 2016b). Countrywide, estimated loss of USD 1.2B for farmers (Agweb, 2016).

Source: Authors' own work based on cited references.

Future precipitation events in the three regions

Table 3.A1.7 provides a review of the influence of ENSO, PDO and IOD indexes on precipitations anomalies, and Table 3.A1.8 matches PDO and ENSO influences on precipitation.

Table 3.A1.7. Influence of positive and negative ENSO, PDO and IOD events on precipitation in the three regions

		India (NW)	United States (California)	China (NE)
ENSO	+	↘	↗	(↘)
	-	↗	↘	(↗)
PDO	+	↘	↘	↘
	-	↗	↗	↗
IOD	+	↗		
	-	↘		

Note: ↘ Favours reduced precipitations; ↗ Favours increased precipitations; ( ) Unclear effect.
Source: Authors, derived from the reviewed literature.

Table 3.A1.8. Combining ENSO and PDO influences on precipitation in the three regions

	Positive ENSO			**Negative ENSO**		
	NW India	SW United States	NE China	NW India	SW United States	NE China
Positive PDO	↘↘*	↗↘	(↘)↘	↗↘*	↘↘	(↗)↘
	Ex: 1997-1998; 2015-2016			Ex: 1988-1989		
Negative PDO	↘↗*	↗↗	(↘)↗	↗↗*	↘↗	(↗)↗
	Ex: 2002-2003			Ex: 2007-2008		

Note: the first arrow indicate the ENSO influence, the second the influence of the PDO.
Source: Authors, derived from the reviewed literature.

Timeline for ENSO, PDO and IOD events

Frequency and duration

- ENSO events last 6-18 months. In the latest 50 years, very strong *El Niño* occur every 15 years approximately (Levrault, 2016), La Nina can occur every 4 to 5 years with very strong La Nina every 23 years (Cai et al., 2015).
- PDO events last 20-30 years, and fluctuate between positive and negative phases.
- IOD fluctuates every 3 to 8 years between positive, negative and neutral phases (IRD, 2013).

According to some studies, if the current global heating trend is maintained, the frequency of ENSO+ events (El Niño) is likely to accelerate, and even double at the end of the century (every 7-8 years) (Levrault, 2016).

Long-range forecasts

The IRI (Columbia University) publishes monthly probabilistic ENSO forecasts up to 10 months ahead. This forecast is the reference used by the NOAA (Climate Prediction Center, 2016). As of 2016, it was considered that there were neutral ENSO conditions. During the fall-winter 2016-2017, La Nina conditions were expected with 75% probability (IRI, 2016).

For the PDO, the predictability relies on the strong inter-seasonal and even inter-annual persistence probability. Models analysing SST-anomalies can predict evolutions of the PDO index three to four seasons in advance (Alexander et al., 2008). PDO shifted in a warm mode in January 2014, so PDO+ conditions are expected to last in the coming years (Browning 2013). IOD long-range forecasts published by the Indian Ministry of Earth Sciences go up to nine months ahead. The latest release (June 2016) anticipates a negative IOD during the 6 next months (Ministry of Earth Sciences, 2016).

Annex 3.A2.

Additional results from the simulations

3.A2.1. Additional results of the simulation for the irrigation stress scenarios

Figure 3.A2.1. Global production and price impact of irrigation stresses under the Hadley and ISPL climate projections by 2050 compared to the baseline

Top: Production, Bottom: World prices, Left: Hadley, Right: ISPL

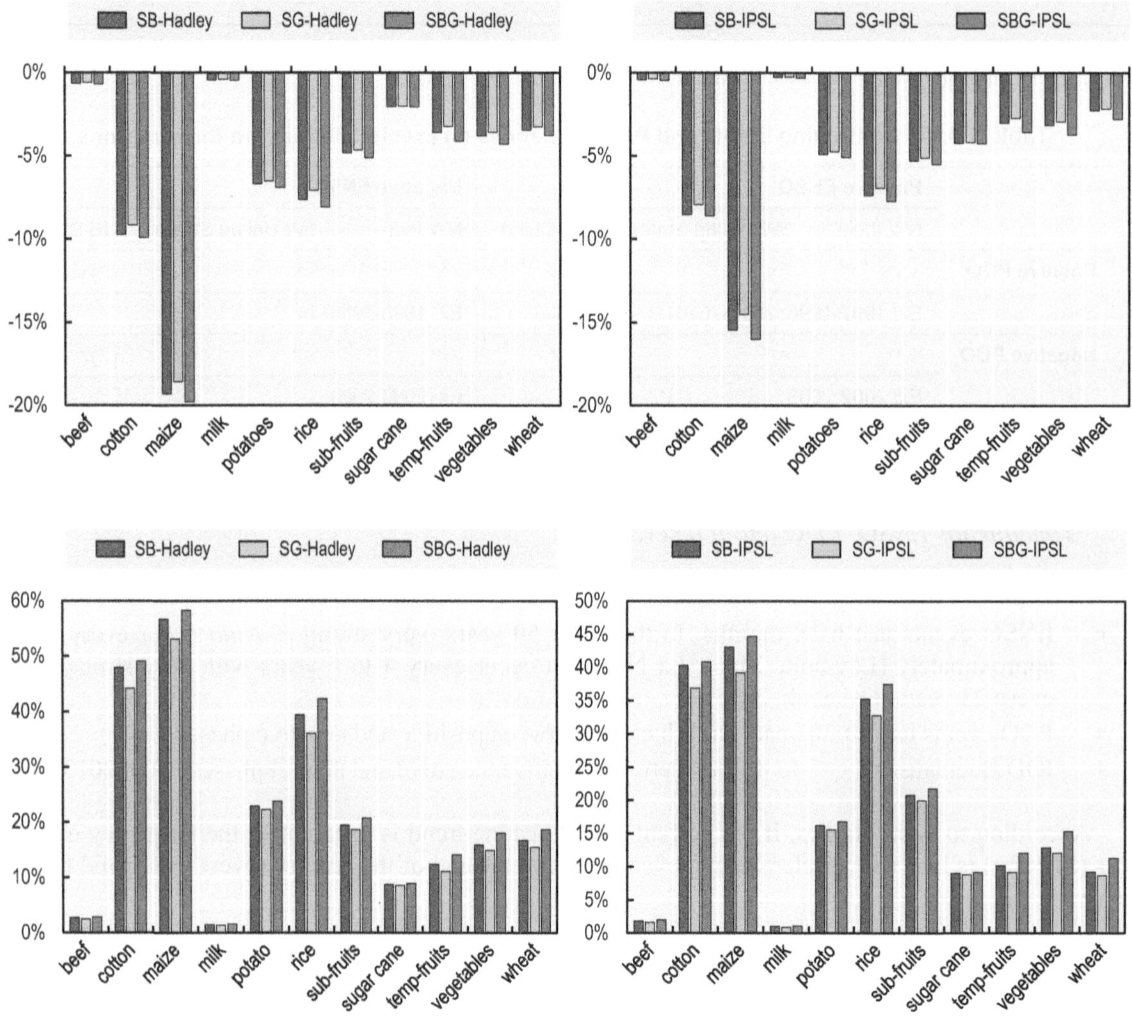

Note: Sub-fruits are subtropical fruits, temp-fruits are temperate fruits.

Source: Derived from IMPACT results.

Figure 3.A2.2. Impact of irrigation stresses and climate change on production in China

Note: Sub-fruits are subtropical fruits, temp-fruits are temperate fruits.
Source: Derived from IMPACT results.

Figure 3.A2.3. Production impact of irrigation stresses and climate change in India

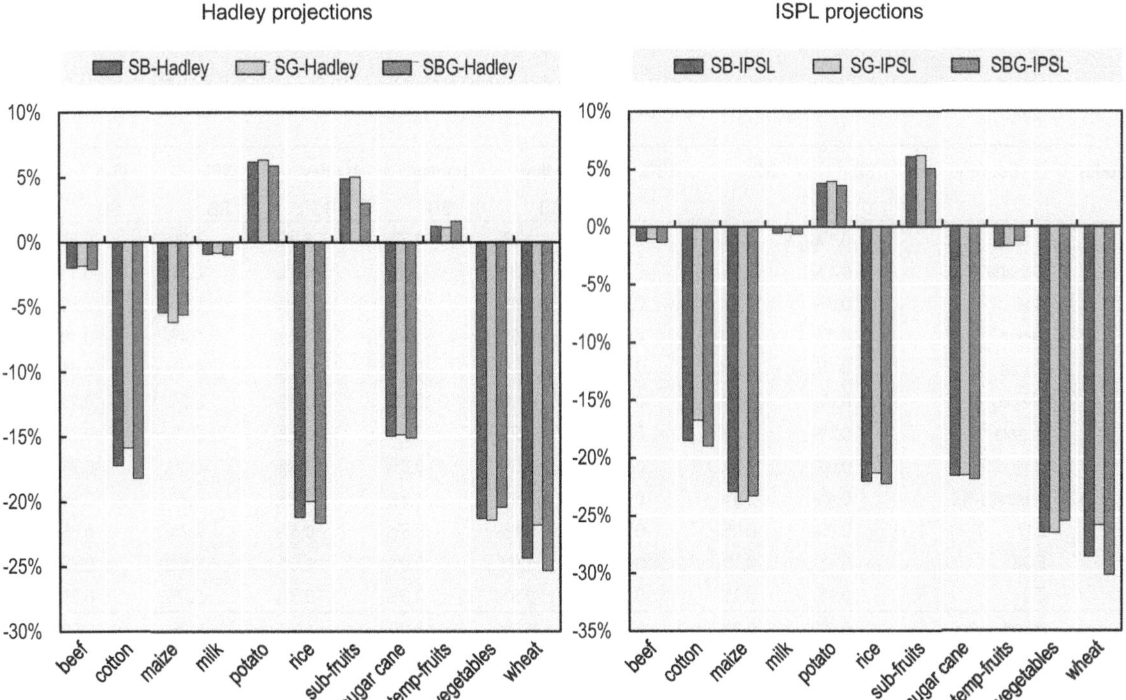

Note: Sub-fruits are subtropical fruits, temp-fruits are temperate fruits.

Source: Derived from IMPACT results.

Figure 3.A2.4. Production impacts of irrigation stresses and climate change in the United States

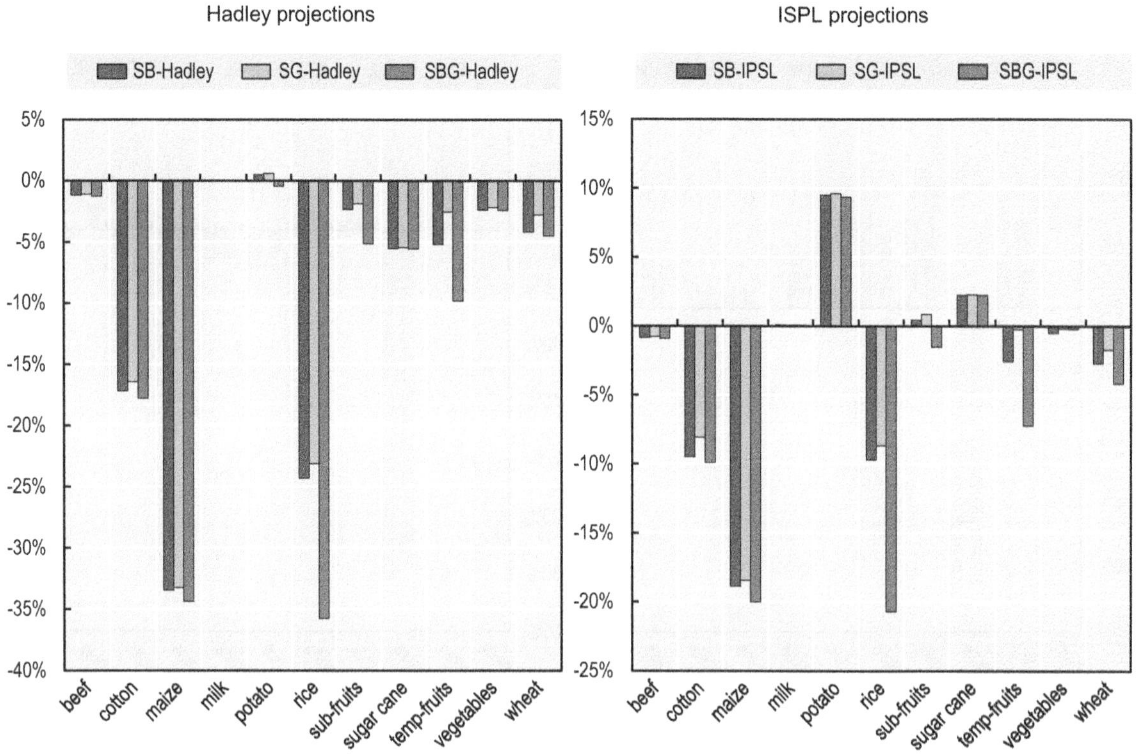

Hadley projections | ISPL projections

SB-Hadley, SG-Hadley, SBG-Hadley | SB-IPSL, SG-IPSL, SBG-IPSL

beef, cotton, maize, milk, potato, rice, sub-fruits, sugar cane, temp-fruits, vegetables, wheat

Note: Sub-fruits are subtropical fruits, temp-fruits are temperate fruits.
Source: Derived from IMPACT results.

Trade impacts

Table 3.A2.1. Relative impact of irrigation stresses on the net trade balance of selected countries

Commodity	Country	SB	SG	SBG	Hadley SB	Hadley SG	Hadley SBG	ISPL SB	ISPL SG	ISPL SBG
Beef	Brazil	-0.1%	0.0%	-0.1%	-1.2%	-1.2%	-1.2%	-0.9%	-0.8%	-0.9%
Beef	Canada	-0.1%	-0.1%	-0.1%	-0.3%	-0.3%	-0.4%	-0.1%	-0.1%	-0.2%
Beef	Chili	0.3%	0.2%	0.3%	1.4%	1.3%	1.5%	0.9%	0.8%	1.0%
Beef	Japan	0.0%	0.0%	0.0%	-0.5%	-0.5%	-0.5%	-0.4%	-0.4%	-0.4%
Beef	Korea	-0.1%	-0.1%	-0.1%	-0.6%	-0.6%	-0.6%	-0.4%	-0.4%	-0.4%
Beef	Mexico	0.0%	0.0%	-0.1%	0.1%	0.1%	0.1%	0.2%	0.2%	0.2%
Beef	Russia	0.0%	0.0%	0.0%	-0.2%	-0.2%	-0.2%	-0.2%	-0.2%	-0.2%
Beef	United States	0.0%	0.0%	0.0%	-0.3%	-0.3%	-0.3%	-0.3%	-0.2%	-0.3%
Beef	Australia-NZ	-0.4%	-0.4%	-0.5%	-1.4%	-1.3%	-1.5%	-0.7%	-0.6%	-0.9%
Beef	E17	0.0%	0.0%	0.0%	0.8%	0.8%	0.8%	0.7%	0.6%	0.7%
Beef	EG4	0.1%	0.1%	0.1%	1.0%	0.9%	1.0%	0.8%	0.7%	0.8%
Beef	EU7	-0.1%	-0.1%	-0.1%	-0.3%	-0.3%	-0.3%	-0.2%	-0.2%	-0.3%
Milk	Brazil	-0.2%	-0.2%	-0.2%	-1.3%	-1.2%	-1.3%	-0.9%	-0.9%	-1.0%
Milk	Canada	0.1%	0.1%	0.1%	0.4%	0.4%	0.5%	0.2%	0.2%	0.3%
Milk	Chili	0.0%	0.0%	0.0%	-0.6%	-0.5%	-0.6%	-0.5%	-0.5%	-0.5%
Milk	Japan	0.0%	0.0%	0.0%	0.0%	0.0%	0.0%	-0.1%	-0.1%	-0.1%
Milk	Korea	0.0%	0.0%	0.0%	0.3%	0.3%	0.3%	0.3%	0.3%	0.3%
Milk	Mexico	0.0%	0.0%	0.1%	0.6%	0.6%	0.6%	0.5%	0.4%	0.5%

Milk	Russia	0.0%	0.0%	0.1%	0.5%	0.4%	0.5%	0.4%	0.3%	0.4%
Milk	United States	0.0%	0.0%	0.0%	0.1%	0.1%	0.1%	0.1%	0.1%	0.1%
Milk	Australia-NZ	0.2%	0.2%	0.3%	0.7%	0.6%	0.7%	0.2%	0.1%	0.3%
Milk	E17	0.1%	0.1%	0.2%	1.1%	1.1%	1.2%	0.8%	0.7%	0.8%
Milk	EG4	0.1%	0.1%	0.2%	1.1%	1.0%	1.1%	0.8%	0.7%	0.8%
Milk	EU7	0.1%	0.0%	0.1%	0.5%	0.5%	0.5%	0.4%	0.3%	0.4%
Maize	Brazil	2.3%	1.3%	2.8%	23.5%	21.8%	24.2%	14.4%	12.2%	15.2%
Maize	Canada	0.3%	-0.4%	0.8%	8.1%	6.9%	8.6%	15.6%	14.1%	16.2%
Maize	Chili	1.5%	1.0%	1.9%	25.1%	24.1%	25.5%	22.7%	21.5%	23.1%
Maize	Japan	-8.7%	-8.5%	-10.5%	-21.5%	-21.1%	-23.2%	-11.8%	-11.5%	-12.7%
Maize	Korea	-3.3%	-3.9%	-3.4%	14.1%	13.3%	14.1%	13.4%	12.4%	13.6%
Maize	Mexico	2.2%	1.5%	2.8%	3.8%	2.5%	4.4%	1.1%	-0.5%	1.8%
Maize	Russia	-9.4%	-10.6%	-11.2%	-78.2%	-79.1%	-79.6%	-28.4%	-30.3%	-28.8%
Maize	United States	2.9%	1.7%	1.3%	-29.9%	-30.3%	-30.6%	-12.4%	-12.6%	-13.4%
Maize	Australia-NZ	2.1%	1.1%	2.5%	31.9%	30.2%	32.5%	21.5%	19.5%	22.2%
Maize	E17	1.1%	0.2%	1.4%	-0.2%	-1.4%	0.1%	9.9%	8.2%	10.4%
Maize	EG4	3%	1%	4%	10.0%	7.8%	11.0%	21.4%	18.5%	22.7%
Maize	EU7	0.8%	0.0%	0.0%	-26.8%	-27.7%	-27.3%	-3.8%	-5.2%	-3.9%
Rice	Brazil	9.6%	8.8%	10.7%	8.7%	7.4%	9.8%	14.5%	13.6%	15.4%
Rice	Canada	1.4%	1.0%	2.1%	11.2%	10.5%	11.9%	10.3%	9.7%	10.8%
Rice	Chili	1.6%	1.0%	2.3%	33.5%	32.4%	34.5%	32.7%	31.9%	33.4%
Rice	Japan	1.1%	0.7%	1.6%	15.4%	14.7%	16.0%	15.5%	14.9%	15.9%
Rice	Korea	1.5%	1.0%	2.1%	8.2%	7.3%	9.0%	6.9%	6.2%	7.5%
Rice	Mexico	1.5%	1.0%	2.3%	6.4%	5.5%	7.1%	4.6%	3.9%	5.2%
Rice	Russia	-0.1%	-1.3%	1.1%	11.7%	10.0%	12.9%	16.8%	15.7%	17.9%
Rice	United States	1.2%	-2.1%	-10.0%	-41.2%	-39.5%	-65.7%	-10.2%	-8.7%	-33.8%
Rice	Australia-NZ	1.9%	1.2%	2.9%	-3.6%	-4.7%	-2.6%	7.8%	6.9%	8.6%
Rice	E17	3.8%	2.6%	5.3%	37.4%	35.3%	39.0%	33.8%	32.3%	35.1%
Rice	EG4	3.3%	2.0%	4.7%	17.5%	15.5%	19.0%	16.5%	15.2%	17.7%
Rice	EU7	1.3%	0.7%	2.1%	14.4%	13.4%	15.2%	17.1%	16.5%	17.8%
Wheat	Brazil	-7.9%	-7.0%	-9.5%	-90.8%	-85.3%	-97.1%	-45.8%	-42.3%	-48.1%
Wheat	Canada	-1.0%	-0.6%	-0.3%	-19.5%	-19.4%	-18.5%	-10.5%	-10.3%	-9.5%
Wheat	Chili	2.5%	2.4%	3.2%	-5.7%	-6.1%	-4.8%	-4.4%	-4.7%	-3.6%
Wheat	Japan	1.4%	1.3%	1.8%	6.5%	6.2%	7.0%	4.9%	4.7%	5.4%
Wheat	Korea	1.4%	1.3%	1.8%	5.3%	5.0%	5.8%	3.3%	3.1%	3.8%
Wheat	Mexico	0.7%	0.7%	0.9%	-3.6%	-3.6%	-3.4%	-2.3%	-2.1%	-1.9%
Wheat	Russia	2.6%	2.3%	4.2%	17.7%	16.7%	19.6%	37.0%	36.0%	38.7%
Wheat	United States	2.4%	2.6%	-0.5%	1.2%	3.1%	1.2%	1.0%	2.3%	-0.9%
Wheat	Australia-NZ	-2.9%	-2.6%	-3.0%	-18.6%	-18.4%	-18.4%	-12.2%	-11.9%	-11.8%
Wheat	E17	3.8%	3.6%	4.7%	4.8%	4.3%	5.7%	4.4%	3.9%	5.3%
Wheat	EG4	5.1%	4.8%	6.4%	0.6%	-0.1%	1.8%	-2.0%	-2.7%	-0.8%
Wheat	EU7	-5.0%	-4.8%	-4.6%	-19.3%	-19.2%	-18.6%	-22.6%	-22.4%	-22.0%
Potatoes	Brazil	1.0%	0.9%	1.4%	1.8%	1.7%	2.1%	7.5%	7.5%	7.7%
Potatoes	Canada	-0.4%	-0.5%	-0.1%	-8.1%	-8.1%	-7.9%	4.5%	4.5%	4.6%
Potatoes	Chili	0.6%	0.4%	0.8%	62.5%	62.5%	62.5%	59.4%	59.5%	59.4%
Potatoes	Japan	0.6%	0.5%	0.7%	2.2%	2.1%	2.3%	-3.5%	-3.6%	-3.4%
Potatoes	Korea	0.8%	0.7%	1.0%	-5.2%	-5.4%	-5.1%	-6.1%	-6.3%	-6.0%
Potatoes	Mexico	0.9%	0.8%	1.4%	11.4%	11.2%	11.7%	-12.7%	-12.9%	-12.5%
Potatoes	Russia	-3.3%	-3.5%	-3.1%	-3.6%	-3.7%	-3.5%	0.4%	0.4%	0.5%
Potatoes	United States	0.5%	0.4%	0.6%	5.9%	5.9%	5.1%	13.8%	13.8%	13.8%
Potatoes	Australia-NZ	0.4%	0.3%	0.7%	4.1%	4.0%	4.2%	3.4%	3.3%	3.6%
Potatoes	E17	-0.3%	-0.4%	-0.1%	-43.1%	-43.2%	-42.9%	-29.6%	-29.6%	-29.5%

Potatoes	EG4	0.3%	0.2%	0.4%	-47.0%	-47.1%	-46.9%	-48.7%	-48.7%	-48.6%
Potatoes	EU7	12.0%	12.0%	12.3%	-23.9%	-23.8%	-23.8%	-3.7%	-3.5%	-3.7%
Vegetables	Brazil	0.7%	0.1%	5.0%	-18.3%	-18.6%	-17.0%	-11.3%	-11.4%	-9.7%
Vegetables	Canada	-16.8%	-17.4%	-13.4%	4.0%	3.5%	5.3%	6.3%	6.1%	8.0%
Vegetables	Chili	0.6%	0.1%	3.7%	29.1%	28.7%	30.2%	30.1%	29.8%	31.4%
Vegetables	Japan	0.2%	-0.2%	2.8%	14.8%	14.5%	15.5%	12.4%	12.3%	13.5%
Vegetables	Korea	0.4%	0.1%	2.7%	3.4%	3.3%	4.1%	3.5%	3.4%	4.4%
Vegetables	Mexico	0.6%	0.0%	4.2%	-8.9%	-9.2%	-7.7%	-10.7%	-10.9%	-9.3%
Vegetables	Russia	-9.1%	-9.8%	-5.7%	-13.0%	-13.4%	-12.1%	-2.2%	-2.4%	-0.8%
Vegetables	United States	-0.4%	-0.8%	2.4%	2.0%	2.0%	2.6%	3.3%	3.4%	4.4%
Vegetables	Australia-NZ	0.5%	0.0%	3.4%	-0.5%	-0.9%	0.4%	2.4%	2.1%	3.6%
Vegetables	E17	0.2%	-0.5%	4.0%	-13.7%	-14.1%	-12.7%	-0.6%	-0.8%	0.9%
Vegetables	EG4	0.3%	-0.3%	3.2%	-7.5%	-7.8%	-6.8%	-10.9%	-11.0%	-9.9%
Vegetables	EU7	-0.3%	-0.8%	3.4%	-18.2%	-18.3%	-17.4%	1.4%	1.6%	2.7%
Temp-fruits	Brazil	1.5%	1.7%	3.3%	-19.7%	-20.0%	-18.6%	-7.1%	-7.0%	-5.7%
Temp-fruits	Canada	-3.5%	-3.5%	-2.7%	7.2%	6.9%	7.8%	7.1%	7.0%	7.8%
Temp-fruits	Chili	2.5%	2.6%	4.6%	53.6%	52.7%	55.3%	56.4%	55.9%	58.3%
Temp-fruits	Japan	1.0%	1.0%	1.7%	3.1%	2.7%	3.6%	2.1%	1.9%	2.7%
Temp-fruits	Korea	1.4%	1.5%	2.4%	5.4%	5.0%	6.2%	4.6%	4.3%	5.5%
Temp-fruits	Mexico	1.8%	1.8%	3.4%	-15.8%	-16.4%	-14.7%	-16.7%	-17.0%	-15.4%
Temp-fruits	Russia	-0.4%	-0.4%	0.7%	5.4%	4.9%	6.1%	8.2%	8.0%	9.1%
Temp-fruits	United States	0.7%	-0.3%	-7.5%	-1.9%	0.4%	-6.2%	0.3%	2.4%	-4.0%
Temp-fruits	Australia-NZ	1.1%	1.1%	2.1%	-3.7%	-4.1%	-3.0%	-0.4%	-0.6%	0.5%
Temp-fruits	E17	1.8%	1.8%	3.1%	-12.3%	-12.7%	-11.5%	-1.8%	-2.0%	-0.8%
Temp-fruits	EG4	1.1%	1.0%	2.1%	-6.9%	-7.2%	-6.2%	-7.1%	-7.2%	-6.2%
Temp-fruits	EU7	0.9%	1.0%	2.3%	-11.2%	-11.5%	-10.4%	0.6%	0.8%	1.7%
Sub-fruits	Brazil	0.0%	0.1%	0.5%	-24.2%	-24.1%	-23.7%	-13.8%	-13.5%	-13.5%
Sub-fruits	Canada	0.5%	0.4%	0.8%	6.1%	5.9%	6.5%	6.7%	6.5%	6.9%
Sub-fruits	Chili	-0.6%	-0.8%	-0.3%	45.5%	45.4%	46.0%	57.1%	57.2%	57.5%
Sub-fruits	Japan	0.2%	0.2%	0.3%	5.1%	5.1%	5.3%	6.3%	6.2%	6.4%
Sub-fruits	Korea	0.6%	0.5%	0.8%	22.2%	21.9%	22.6%	19.0%	18.7%	19.2%
Sub-fruits	Mexico	-0.1%	0.0%	0.3%	-13.7%	-13.6%	-13.3%	-22.0%	-21.9%	-21.8%
Sub-fruits	Russia	-1.5%	-1.6%	-1.3%	9.1%	9.0%	9.3%	13.0%	12.9%	13.1%
Sub-fruits	United States	0.1%	-0.1%	-0.6%	3.3%	3.5%	0.9%	6.5%	6.7%	4.9%
Sub-fruits	Australia-NZ	0.0%	-0.2%	0.3%	2.4%	2.2%	2.7%	2.9%	2.7%	3.1%
Sub-fruits	E17	1.0%	0.9%	1.3%	-9.1%	-9.2%	-8.7%	-4.4%	-4.6%	-4.2%
Sub-fruits	EG4	0.4%	0.3%	0.6%	-0.4%	-0.6%	-0.1%	-1.6%	-1.7%	-1.4%
Sub-fruits	EU7	0.1%	0.1%	0.4%	-8.3%	-8.2%	-7.9%	6.1%	6.4%	6.3%
Sugarcane	Brazil	0.9%	0.8%	1.1%	8.8%	8.7%	8.9%	10.7%	10.6%	10.9%
Sugarcane	Canada	0.2%	0.1%	0.2%	1.9%	1.9%	1.9%	2.3%	2.3%	2.3%
Sugarcane	Chili	0.6%	0.5%	0.7%	9.9%	9.8%	10.0%	10.7%	10.6%	10.8%
Sugarcane	Japan	0.3%	0.2%	0.2%	5.0%	5.0%	4.9%	5.8%	5.8%	5.8%
Sugarcane	Korea	0.3%	0.2%	0.3%	3.4%	3.4%	3.5%	3.6%	3.5%	3.6%
Sugarcane	Mexico	0.7%	0.6%	0.8%	-6.9%	-6.8%	-6.8%	-9.1%	-9.1%	-9.1%
Sugarcane	Russia	-0.8%	-0.9%	-0.8%	4.0%	3.9%	4.0%	9.1%	9.0%	9.1%
Sugarcane	United States	0.1%	0.0%	-0.2%	1.6%	1.8%	1.4%	7.5%	7.5%	7.3%
Sugarcane	Australia-NZ	0.2%	0.0%	0.0%	8.7%	8.8%	8.6%	9.8%	9.9%	9.7%
Sugarcane	E17	-0.7%	-0.9%	-0.8%	-10.5%	-10.5%	-10.6%	5.5%	5.5%	5.6%
Sugarcane	EG4	0.2%	0.1%	0.0%	-16.1%	-16.0%	-16.4%	0.5%	0.7%	0.4%
Sugarcane	EU7	0.9%	0.8%	0.9%	-2.0%	-2.0%	-2.1%	6.5%	6.5%	6.5%
Cotton	Brazil	5.2%	3.9%	7.0%	-25.7%	-28.3%	-25.1%	-13.0%	-15.0%	-12.5%
Cotton	Canada	1.4%	1.1%	1.9%	11.1%	10.4%	11.3%	9.7%	9.0%	9.8%

Cotton	Chili	0.3%	0.3%	0.4%	2.7%	2.5%	2.8%	2.3%	2.2%	2.4%
Cotton	Japan	0.8%	0.6%	1.0%	6.3%	5.9%	6.4%	5.4%	5.0%	5.5%
Cotton	Korea	1.4%	1.1%	1.9%	11.1%	10.4%	11.3%	9.7%	9.0%	9.8%
Cotton	Mexico	1.0%	0.7%	1.3%	2.2%	1.5%	2.3%	1.4%	0.8%	1.5%
Cotton	Russia	1.4%	1.1%	1.9%	11.1%	10.4%	11.3%	9.7%	9.0%	9.8%
Cotton	United States	-0.3%	-4.2%	-10.9%	-40.9%	-39.3%	-42.5%	-19.1%	-15.5%	-20.3%
Cotton	Australia-NZ	83.0%	48.8%	109.9%	-778.7%	-843.5%	-769.2%	181.4%	121.7%	194.0%
Cotton	E17	17.7%	14.8%	20.0%	22.5%	16.9%	23.2%	20.6%	16.1%	21.4%
Cotton	EG4	1.4%	1.1%	1.9%	11.1%	10.4%	11.3%	9.7%	9.0%	9.8%
Cotton	EU7	1.5%	1.1%	2.0%	10.7%	9.9%	10.9%	9.3%	8.6%	9.4%

1. The absolute size of trade change depends on the initial volume of trade flows, net trade changes can result in shift in the direction of trade flows, which should be accounted for when interpreting these results.

2. EG4: France, Germany, Italy, United Kingdom, E17: Austria, Belgium, Luxemburg, Netherlands, Ireland, Spain, Portugal, Greece, Czech Republic, Slovakia, Poland, Hungary, Estonia, Finland, Denmark, Sweden Slovenia; EU7: Bulgaria, Lithuania, Latvia, Croatia, Malta, Cyprus, Romania.

3. Sub-fruits are subtropical fruits, temp-fruits are temperate fruits.

Source: Derived from IMPACT simulations.

3.A2.2. Results of the simulation for the drought scenarios

Figure 3.A2.5. Changes in production compared to the baseline for China (left) and India (right) with a drought and climate change scenarios in the United States in 2021

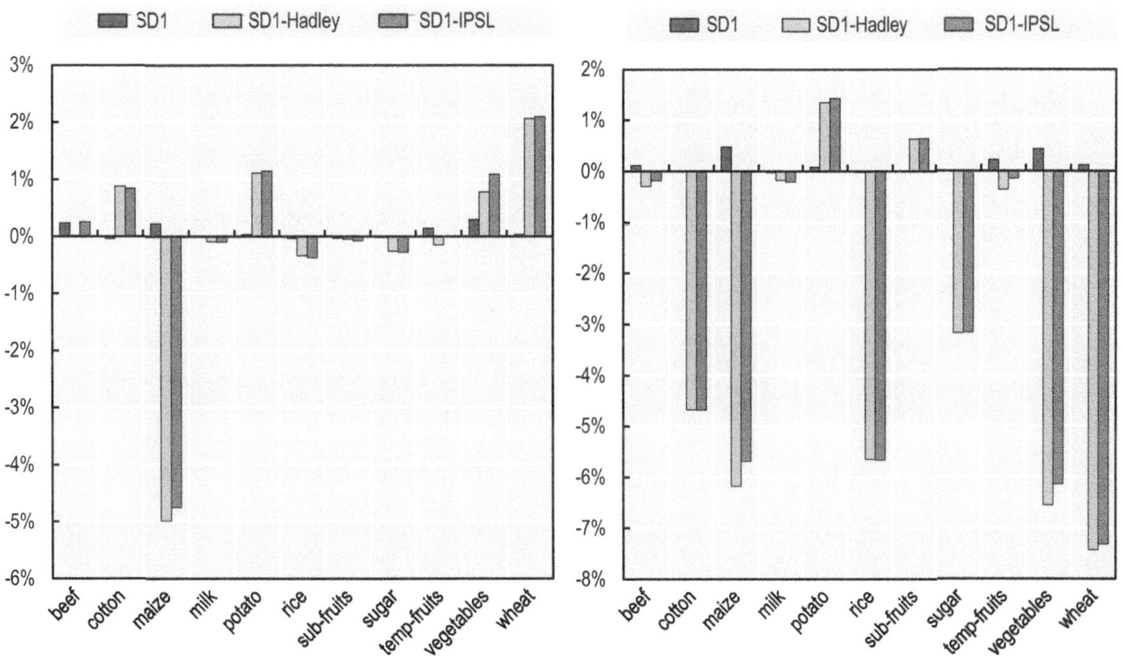

Note: Sub-fruits are subtropical fruits, temp-fruits are temperate fruits.
Source: Derived from IMPACT results.

Figure 3.A2.6. Changes in US production compared to the baseline with a drought and climate change scenarios in China and India in 2030

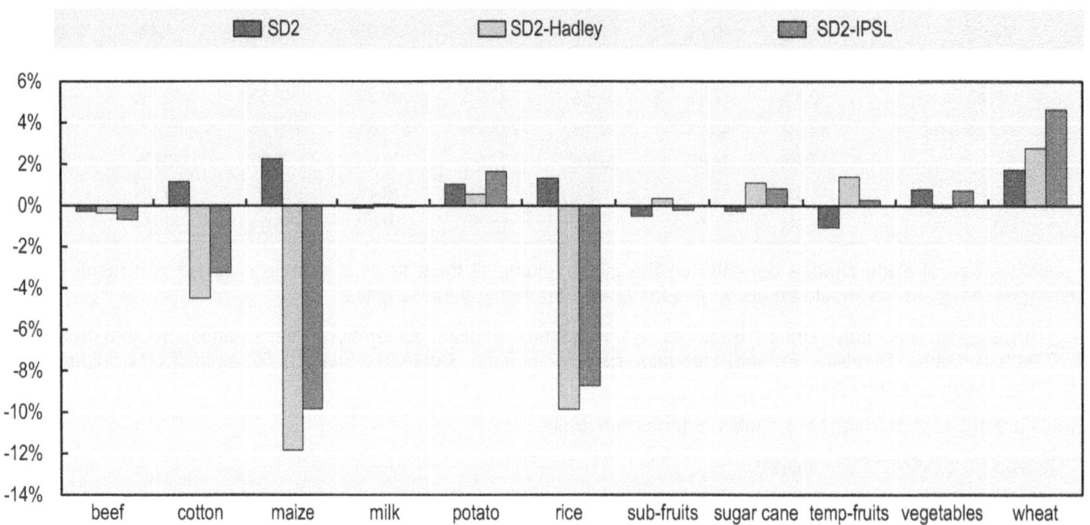

Note: Sub-fruits are subtropical fruits, temp-fruits are temperate fruits.

Source: Derived from IMPACT results.

Trade effects of the two drought scenarios

Table 1. Table 3.A2.2. Effects of the drought scenarios on net trade

Crop	Country	SD1	SD1-Hadley	SD1-IPSL	SD2	SD2-Hadley	SD2-IPSL
Beef	Brazil	0.9%	0.6%	0.7%	0.0%	-0.5%	-0.4%
Beef	Canada	0.7%	0.6%	0.7%	-0.2%	-0.3%	-0.2%
Beef	Chile	0.6%	0.8%	0.7%	0.4%	0.8%	0.7%
Beef	China	0.5%	0.6%	0.6%	0.0%	0.2%	0.2%
Beef	E17	0.6%	0.8%	0.8%	0.0%	0.4%	0.3%
Beef	EG4	0.6%	0.8%	0.7%	0.1%	0.5%	0.4%
Beef	EU7	0.4%	0.4%	0.4%	-0.1%	-0.1%	-0.1%
Beef	India	0.5%	0.3%	0.4%	-0.5%	-0.9%	-0.8%
Beef	Japan	0.4%	0.3%	0.3%	0.0%	-0.1%	-0.1%
Beef	Korea	0.2%	0.1%	0.1%	-0.1%	-0.2%	-0.2%
Beef	Mexico	0.5%	0.6%	0.6%	0.0%	0.2%	0.2%
Beef	Australia-NZ	1.2%	0.9%	1.0%	-0.6%	-1.0%	-0.9%
Beef	Russia	0.4%	0.4%	0.4%	-0.1%	-0.2%	-0.2%
Beef	United States	-2.7%	-2.8%	-2.8%	-0.1%	-0.2%	-0.2%
Beef	South Africa	0.6%	0.5%	0.6%	-0.1%	-0.1%	-0.1%
Cotton	Brazil	-0.2%	-4.7%	-4.3%	6.2%	-2.8%	-2.0%
Cotton	Canada	0.0%	2.6%	2.4%	1.9%	6.6%	6.2%
Cotton	Chile	0.0%	0.6%	0.6%	0.5%	1.6%	1.5%
Cotton	China	0.1%	4.8%	4.4%	-4.9%	3.3%	2.7%
Cotton	E17	-0.2%	5.8%	4.6%	9.9%	21.8%	19.1%
Cotton	EG4	0.0%	2.6%	2.4%	1.9%	6.6%	6.2%
Cotton	EU7	0.0%	2.6%	2.4%	2.0%	6.6%	6.2%
Cotton	India	0.0%	-3.4%	-4.1%	1.7%	-3.3%	-4.3%
Cotton	Japan	0.0%	1.4%	1.3%	1.1%	3.7%	3.5%
Cotton	Korea	0.0%	2.6%	2.4%	1.9%	6.6%	6.2%
Cotton	Mexico	0.0%	0.3%	0.2%	1.5%	2.2%	2.1%
Cotton	Australia-NZ	-1.8%	-237.1%	32.9%	143.4%	-276.5%	216.8%
Cotton	Russia	0.0%	2.6%	2.4%	1.9%	6.6%	6.2%
Cotton	United States	-0.2%	-4.1%	-0.7%	5.1%	-2.3%	4.1%
Cotton	South Africa	-0.2%	1.9%	2.0%	3.8%	7.7%	7.8%
Maize	Brazil	0.6%	5.7%	2.5%	4.3%	13.9%	8.0%
Maize	Canada	0.4%	2.4%	4.7%	3.0%	6.4%	10.0%
Maize	Chile	0.3%	5.8%	5.7%	2.4%	12.7%	12.3%
Maize	China	1.0%	4.3%	1.0%	-20.0%	-12.8%	-17.5%
Maize	E17	0.6%	2.6%	3.7%	3.5%	6.4%	8.4%
Maize	EG4	0.8%	6.4%	7.3%	7%	16%	18%
Maize	EU7	0.4%	-1.6%	-0.8%	1.1%	-2.5%	-1.3%
Maize	India	0.5%	-3.3%	-5.9%	1.9%	-3.7%	-7.5%
Maize	Japan	0.0%	-3.5%	-2.8%	-1.5%	-8.1%	-6.7%
Maize	Korea	0.4%	3.8%	3.6%	2.1%	8.4%	7.9%
Maize	Mexico	0.5%	0.2%	1.8%	3.7%	3.0%	6.1%
Maize	Australia-NZ	0.5%	7.1%	6.3%	3.4%	15.7%	14.1%
Maize	Russia	0.3%	-6.3%	-4.6%	-0.9%	-14.8%	-11.4%
Maize	United States	-1.5%	-5.6%	-2.4%	6.6%	-2.0%	3.4%
Maize	South Africa	0.4%	0.1%	-3.5%	3.0%	2.8%	-3.1%
Milk	Brazil	0.0%	-0.3%	-0.2%	-0.3%	-0.7%	-0.7%
Milk	Canada	0.0%	0.1%	0.0%	0.1%	0.2%	0.2%
Milk	Chile	0.0%	-0.2%	-0.2%	-0.1%	-0.3%	-0.3%
Milk	China	0.0%	0.0%	0.0%	0.1%	0.1%	0.1%
Milk	E17	0.0%	0.2%	0.2%	0.2%	0.6%	0.5%
Milk	EG4	0.0%	0.2%	0.2%	0.2%	0.6%	0.5%
Milk	EU7	0.0%	0.1%	0.1%	0.1%	0.3%	0.2%
Milk	India	0.0%	-0.1%	-0.1%	0.0%	-0.2%	-0.1%
Milk	Japan	0.0%	0.0%	0.0%	0.0%	0.0%	0.0%
Milk	Korea	0.0%	0.0%	0.0%	0.0%	0.1%	0.1%
Milk	Mexico	0.0%	0.1%	0.1%	0.0%	0.3%	0.2%
Milk	Australia-NZ	0.0%	0.1%	0.0%	0.2%	0.3%	0.2%
Milk	Russia	0.0%	0.1%	0.1%	0.1%	0.2%	0.2%
Milk	United States	0.0%	0.0%	0.0%	-0.1%	0.0%	0.0%

Milk	South Africa	0.0%	0.0%	0.0%	0.1%	0.2%	0.1%
Potatoes	Brazil	0.1%	0.6%	1.9%	3.1%	3.7%	6.1%
Potatoes	Canada	0.1%	0.3%	3.4%	3.1%	3.1%	8.5%
Potatoes	Chile	0.1%	13.7%	12.7%	3.4%	30.2%	28.1%
Potatoes	China	0.1%	3.3%	2.0%	-11.2%	-6.1%	-8.4%
Potatoes	E17	0.1%	-11.8%	-7.3%	2.6%	-18.8%	-10.5%
Potatoes	EG4	0%	-13%	-13%	2.0%	-22.0%	-22.4%
Potatoes	EU7	0.0%	-3.2%	-1.1%	2.8%	-4.4%	-0.2%
Potatoes	India	0.2%	3.4%	2.4%	2.5%	7.8%	6.1%
Potatoes	Japan	0.1%	0.8%	-0.6%	1.9%	2.8%	0.3%
Potatoes	Korea	0.1%	-1.1%	-1.2%	2.3%	-0.2%	-0.4%
Potatoes	Mexico	0.1%	2.9%	-3.6%	3.0%	7.8%	-3.8%
Potatoes	Australia-NZ	0.1%	2.7%	2.1%	2.1%	6.3%	5.3%
Potatoes	Russia	0.1%	3.5%	3.1%	2.5%	8.4%	7.7%
Potatoes	United States	-1.9%	-0.3%	1.5%	2.4%	5.4%	8.9%
Potatoes	South Africa	0.0%	2.0%	1.3%	2.4%	5.9%	4.5%
Rice	Brazil	0.0%	0.3%	1.5%	3.6%	4.2%	6.4%
Rice	Canada	0.0%	2.9%	2.9%	2.3%	7.4%	7.2%
Rice	Chile	0.0%	8.4%	8.4%	2.9%	18.7%	18.5%
Rice	China	0.0%	2.1%	2.1%	-4.4%	-0.9%	-1.0%
Rice	E17	0.0%	7.9%	7.2%	4.5%	19.7%	18.2%
Rice	EG4	-0.1%	5.1%	4.7%	4.4%	13.9%	12.9%
Rice	EU7	0.0%	3.5%	3.7%	2.2%	8.9%	9.3%
Rice	India	0.0%	-3.8%	-4.3%	-2.4%	-8.7%	-9.2%
Rice	Japan	0.0%	3.4%	3.6%	1.8%	8.3%	8.5%
Rice	Korea	0.0%	1.7%	1.6%	2.1%	5.3%	5.0%
Rice	Mexico	0.0%	1.8%	1.9%	2.6%	5.8%	5.8%
Rice	Australia-NZ	0.0%	-2.4%	2.0%	3.6%	-0.1%	7.0%
Rice	Russia	0.0%	3.8%	4.3%	3.3%	10.7%	11.6%
Rice	United States	-0.1%	-7.3%	0.4%	5.0%	-8.4%	5.5%
Rice	South Africa	0.0%	2.9%	2.8%	2.3%	7.4%	7.1%
Sub-fruits	Brazil	-2.4%	-148.0%	-112.7%	-0.4%	-10.5%	-8.1%
Sub-fruits	Canada	0.1%	3.4%	3.2%	0.6%	3.5%	3.3%
Sub-fruits	Chile	0.0%	5.2%	5.4%	-1.3%	24.7%	25.7%
Sub-fruits	China	1.1%	124.7%	114.9%	0.4%	2.6%	2.4%
Sub-fruits	E17	-0.1%	1.8%	-6.9%	-0.1%	0.0%	-1.8%
Sub-fruits	EG4	-0.1%	4.2%	-8.6%	0.1%	0.6%	-1.2%
Sub-fruits	EU7	-0.1%	-0.2%	1.7%	-1.0%	-1.4%	3.2%
Sub-fruits	India	1.9%	82.8%	75.0%	-0.5%	2.3%	2.0%
Sub-fruits	Japan	0.0%	4.4%	3.9%	-0.1%	2.3%	2.0%
Sub-fruits	Korea	0.0%	5.2%	5.3%	0.5%	5.1%	5.2%
Sub-fruits	Mexico	-1.9%	-55.3%	-51.0%	-0.4%	-7.0%	-6.5%
Sub-fruits	Australia-NZ	0.0%	3.3%	3.1%	-0.1%	4.7%	4.5%
Sub-fruits	Russia	0.0%	4.2%	4.1%	-0.4%	3.1%	3.0%
Sub-fruits	United States	0.0%	40.2%	37.3%	0.2%	3.0%	2.8%
Sub-fruits	South Africa	-0.2%	-2.8%	3.4%	-0.8%	-6.3%	5.2%
Temp-fruits	Brazil	0.4%	-3.9%	-1.2%	-0.5%	-9.0%	-3.6%
Temp-fruits	Canada	0.2%	2.3%	2.3%	0.0%	4.0%	4.1%
Temp-fruits	Chile	0.6%	14.1%	14.8%	0.6%	25.3%	26.7%
Temp-fruits	China	0.4%	1.0%	0.9%	0.4%	1.3%	1.2%
Temp-fruits	E17	0.4%	-2.1%	-0.2%	-0.3%	-5.1%	-1.5%
Temp-fruits	EG4	0.3%	-0.9%	-1.8%	-0.4%	-2.7%	-4.2%
Temp-fruits	EU7	0.3%	-0.4%	1.0%	-1.1%	-2.7%	0.0%
Temp-fruits	India	0.5%	1.1%	1.0%	0.6%	1.8%	1.5%
Temp-fruits	Japan	0.3%	0.5%	0.4%	0.1%	0.4%	0.3%
Temp-fruits	Korea	0.4%	0.9%	1.0%	0.3%	1.1%	1.3%
Temp-fruits	Mexico	0.5%	-3.9%	-4.1%	0.3%	-8.1%	-8.4%
Temp-fruits	Australia-NZ	0.3%	-1.5%	-0.2%	0.3%	-2.8%	-0.7%
Temp-fruits	Russia	0.3%	1.7%	1.9%	-0.1%	2.3%	2.6%
Temp-fruits	United States	-5.0%	-3.7%	-3.7%	-0.5%	1.9%	2.0%
Temp-fruits	South Africa	0.3%	-1.3%	-1.2%	-1.3%	-4.3%	-4.1%
Sugarcane	Brazil	0.0%	1.6%	2.1%	0.1%	2.8%	3.6%
Sugarcane	Canada	0.0%	0.4%	0.6%	0.0%	0.8%	1.2%
Sugarcane	Chile	0.0%	2.2%	2.5%	0.1%	4.5%	5.3%
Sugarcane	China	0.0%	0.4%	0.7%	0.0%	0.7%	1.1%
Sugarcane	E17	0.0%	0.5%	2.0%	-0.4%	0.1%	3.4%
Sugarcane	EG4	0.0%	0.2%	1.1%	-0.8%	-1.2%	0.9%
Sugarcane	EU7	0.0%	0.4%	1.4%	-0.2%	0.6%	2.7%

Sugarcane	India	0.0%	-1.8%	-2.7%	0.1%	-3.2%	-5.1%
Sugarcane	Japan	0.0%	0.8%	1.1%	-0.2%	1.5%	2.0%
Sugarcane	Korea	0.0%	0.7%	0.9%	0.2%	1.5%	1.8%
Sugarcane	Mexico	0.0%	-1.8%	-1.0%	0.0%	-3.7%	-2.0%
Sugarcane	Australia-NZ	0.0%	3.4%	4.4%	-1.0%	5.3%	7.3%
Sugarcane	Russia	0.0%	1.5%	1.9%	0.0%	3.1%	3.9%
Sugarcane	United States	0.0%	1.2%	1.7%	-0.2%	2.3%	3.3%
Sugarcane	South Africa	0.0%	1.2%	0.2%	-0.2%	2.6%	0.7%
Vegetables	Brazil	0.4%	-3.2%	-2.1%	3.4%	-3.5%	-1.6%
Vegetables	Canada	0.3%	6.0%	6.1%	2.8%	14.5%	14.8%
Vegetables	Chile	0.3%	7.6%	8.0%	2.9%	17.0%	17.7%
Vegetables	China	0.5%	2.2%	2.2%	-4.4%	-1.1%	-1.2%
Vegetables	E17	0.4%	-0.9%	0.8%	3.1%	0.6%	3.9%
Vegetables	EG4	0.3%	-1.2%	-2.4%	2.1%	-0.8%	-2.8%
Vegetables	EU7	0.3%	1.6%	3.0%	2.0%	4.4%	7.1%
Vegetables	India	0.6%	-4.4%	-5.8%	4.5%	-3.8%	-6.1%
Vegetables	Japan	0.3%	2.2%	2.4%	1.8%	5.7%	6.1%
Vegetables	Korea	0.3%	1.1%	1.3%	1.8%	3.3%	3.7%
Vegetables	Mexico	0.4%	-2.6%	-2.6%	3.3%	-1.8%	-1.8%
Vegetables	Australia-NZ	0.4%	0.2%	1.3%	2.8%	2.7%	4.6%
Vegetables	Russia	0.4%	3.9%	4.3%	2.2%	9.3%	10.1%
Vegetables	United States	-9.2%	-8.3%	-7.7%	2.6%	4.7%	5.6%
Vegetables	South Africa	0.3%	-1.1%	-0.5%	2.8%	-0.1%	1.1%
Wheat	Brazil	-0.1%	-8.2%	-5.8%	0.0%	-20.8%	-15.0%
Wheat	Canada	0.2%	-2.4%	-1.8%	5.8%	1.3%	2.3%
Wheat	Chile	0.1%	-1.7%	-1.0%	3.6%	0.1%	1.4%
Wheat	China	0.2%	2.7%	3.3%	-18.1%	-14.0%	-13.4%
Wheat	E-17	0.3%	1.9%	1.7%	5.1%	7.3%	7.0%
Wheat	EG4	0.4%	-0.4%	-0.2%	6.3%	4.4%	4.8%
Wheat	EU7	0.0%	-2.4%	-2.1%	4.1%	-0.6%	0.1%
Wheat	India	0.3%	-5.9%	-6.9%	-2.5%	-12.1%	-13.8%
Wheat	Japan	0.1%	1.1%	1.1%	2.1%	3.9%	4.0%
Wheat	Korea	0.1%	0.7%	0.5%	2.1%	3.2%	3.0%
Wheat	Mexico	0.1%	-1.4%	-0.8%	2.3%	-0.5%	0.7%
Wheat	Australia-NZ	0.1%	-3.1%	-2.5%	3.4%	-2.4%	-1.4%
Wheat	Russia	0.3%	5.8%	6.1%	6.8%	18.5%	19.3%
Wheat	United States	-4.2%	0.0%	-1.8%	5.7%	13.2%	10.1%
Wheat	South Africa	0.1%	-2.7%	-2.1%	2.8%	-2.1%	-0.9%

1. The absolute size of trade change depends on the initial volume of trade flows, net trade changes can result in shift in the direction of trade flows, which should be accounted for when interpreting these results.

2. EG4: France, Germany, Italy, United Kingdom, E17: Austria, Belgium, Luxemburg, Netherlands, Ireland, Spain, Portugal, Greece, Czech Republic, Slovakia, Poland, Hungary, Estonia, Finland, Denmark, Sweden Slovenia; EU7: Bulgaria, Lithuania, Latvia, Croatia, Malta, Cyprus, Romania.

3. Sub-fruits are subtropical fruits, temp-fruits are temperate fruits.

Source: Results from the IMPACT simulations.

Notes

1. Andrews et al. (2004) also find that for a given exceedance probability, the ratio of the El Niño to non–El Niño annual peak floods varies from more than 10 near 32°N to less than 0.7 near 42°N.

Chapter 4

Confronting future water risks

This chapter discusses policy responses that seek to reduce water risks in hotspots. An economic model is used to assess the role of farmers, private companies, and governments in mitigating water risks. A policy action plan is proposed to address water risks for agriculture production in hotspot locations, mitigate the market impacts that may result, and alleviate broader socio-economic impacts.

The statistical data for Israel are supplied by and under the responsibility of the relevant Israeli authorities. The use of such data by the OECD is without prejudice to the status of the Golan Heights, East Jerusalem and Israeli settlements in the West Bank under the terms of international law.

Key messages

Farmers, agro-food companies and governments share the burden and responsibility of confronting future water risks in hotspot areas, and will respond differently according to their incentives and capacity to respond. Farmers will respond to future water risks if sufficiently well informed of impacts and solutions, by adopting better practices or by shifting activities. However, they may not be as responsive to water risks to which they contribute. For instance, a farmer is more likely to change practices in response to a drought than reducing his use of fertilisers or pesticides in response to a pollution problem.

Agro-food companies in hotspot regions face different constraints and are more likely to respond when they heavily rely on products vulnerable to water risks. Companies may resort to purchasing products from farms outside hotspot regions, they may change their agricultural inputs or ingredients, or may partner with farmers to address risks under stewardship programmes. Large companies, in particular, may have more incentive to respond to agriculture water risks when they face reputational damage.

Governments should complement efforts by private stakeholders to mitigate water risks and co-ordinate reponses in hotspot regions. Governments should incentivise farmers to engage in risk reducing behaviours by increasing the benefits of acting or the cost of inaction and encouraging supply chain actors to also play their part.

This chapter proposes a three-tier action plan for governments that incorporates the hotspots approach.

1. At the hotspot level, governments should adapt existing national policy instruments, such as a cap and trade water systems, to conditions in hotspot regions, and introduce targeted instruments directly addressing water risks, such as locally-tailored information provision and extension efforts.

2. At the national and international level, governments should ensure that markets are well integrated and functioning so that when a problem arises it does not have a ripple effect. Well-functioning food markets play a key role in diluting price effects and ensuring that regional effects are contained. International trade partners should ensure that markets remain open to avoid having market reactions worsening the problem.

3. At a broader level, international collaboration aimed at increasing the resilience of vulnerable hotspot regions can mitigate risks, including through information and technology sharing and financial assistance. For example, China and the western states of the United States have collaborated with the Australian government to learn about the functioning of water markets in the Murray Darling Basin. Countries not facing water risks themselves should anticipate how water shocks in other countries can affect them through food insecurity, social unrest or migration.

The medium-term outlook suggests that the required policy actions will be different in each of the three hotspots studied. The Southwest United States should continue to vigourously apply policies already in place, but agriculture will need to adapt to drier conditions. Northeast China would need to reinforce and scale-up current policy initiatives, such as the use of water pricing to curb water demand, and decouple agricultural support from input use. Northwest India would need to overhaul existing policies such as energy subsidies and the promotion of solar pumps that may accelerate groundwater depletion. Steps already taken, including the introduction of a groundwater model bill, suggest a willingness to move in the right direction.

4.1. From project impacts to effective responses: Who, what and how to confront future water risks?

In the previous chapters, the hotspot approach has been proposed and used to identify countries and regions subject to future agriculture water risks, the possible impacts they may have on the agro-food system and beyond have then been assessed. This chapter analyses and identifies solutions to respond to such risks. The proposed responses will be assessed generally and are applied differentially in the three identified hotspot regions and in others projected to face high agriculture water risks.

Three questions are addressed. First, who can and should respond to acute water risks? Given that high water risks generate private and public damages, governments should share the responsibilities with farmers and companies. What should they do? A simple economic model will be used to assess the incentives farmers and companies in hotspot locations face to respond to water risks, to understand what actions they are likely to undertake, and consequently, to identify what governments should do and in which circumstances. A policy plan is then proposed to respond to the three layers of impacts defined in Chapter 3: production impacts in hotspot regions, direct market impacts in other regions and countries and indirect food security impacts.

The chapter is organised in four subsections. Section 4.2 discusses what the role of farmers and food companies may be in addressing water risks in hotspot regions. Section 4.3 presents the proposed action plan for governments. The plan is then detailed in two parts; Section 4.4 outlines policy options to mitigate agriculture production risks at the hotspot location, and Section 4.5 describes measures to reduce broader sector risks, and to limit chain reaction that would affect non-hotspot countries. Section 4.6 closes the chapter with a discussion of implications for three pre-identified hotspot regions.

4.2. Sharing the burden and responsibilities of responding to acute water risks: The role of farmers, food companies and governments

Just as water risk impacts can differently affect producers and other agro-food market participants in hotspot regions (Chapters 1 to 3), they each have different motives to respond to water risks. This section analyses the incentives for reaction and possible responses of farmers and agro-purchasing food companies in hotspot regions. It addresses ongoing and future water risks and identifies what complementary role governments may play in that setting.

As a preamble, an important distinction should be made between: exogenous and endogenous risks. Water risks that farmers face exogenously—which will be denoted as exogenous risks—including those related to climate change and competing water demand on which an individual farmer (and food company) has limited effect. In contrast, endogenous water risks are risks to which an individual farmer (or company) contributes, e.g. nutrient runoffs leading to water pollution or intensive groundwater use leading to aquifer depletion. Even if most situations will involve a mix of both risks, the set of incentives farmers and food companies face may differ accordingly.

A simple economic model can help identify the critical factors affecting the likelihood and types of responses to water risks (the details of the derivations are presented in Annex 4.A1).[1] Assume that an economic agent (farmer or agro-food company) is confronted with two possible scenarios, it may be facing acute future water risks with probability φ, or no water risks with probability $(1-\varphi)$. The expected profits the agent faces can be expressed as $E(\Pi/NR) = \varphi.\Pi_1 + (1-\varphi).\overline{\Pi_1}$, assuming Π_1 is the profit under water risks and $\overline{\Pi_1}$ is the agent's profit under no risk both at time t1 (future) with no reaction (NR).[2] A similar expression can be derived if the agent decides to react to this possibility (noted R): $E(\Pi/R) = \varphi.\Pi_R + (1-\varphi).\overline{\Pi_R}$ with Π_R and $\overline{\Pi_R}$ as the profit derived with this reaction under risks and no risks, respectively. A perfectly informed agent will then respond to future water risks if and only if: $E(\Pi/R) \geq E(\Pi/NR)$ which is equivalent to the following arbitrage equation:

$$\varphi[(\overline{\Pi_1} - \Pi_1) - (\overline{\Pi_R} - \Pi_R)] \geq \overline{\Pi_1} - \overline{\Pi_R} \quad (1)$$

This equation can be interpreted as an incentive compatibility constraint. It means that the agent will only take action if the net benefit from reaction under risks exceeds the costs it involves under no risks. The probability θ that equation (1) is verified, which can be interpreted as the likelihood of reaction, increases with the probability of water risks (φ) and the expected profit losses with inaction if such risk were realised ($\overline{\Pi_1} - \Pi_1$). Conversely, this probability decreases with the remaining water risk impacts with the response ($\overline{\Pi_R} - \Pi_R$) and the relative costs the response may involve in a case of the absence of water risks ($\overline{\Pi_1} - \overline{\Pi_R}$).

Naturally, the profit functions used in (1) will differ according to different actors, and whether agents face exogenous risks or directly affect future water risks (endogenous risks). The next two sub-sections will use this arbitrage equation as a basis to discuss responses by farmers and companies.

Farmers respond more proactively to short- or medium-term exogenous water risks by changing practices and technology or by switching activities

The responses of individual farmers to water risks in the hotspot region are subject to multiple factors. In the immediate or medium term, farmers with information and access to capital will take action to mitigate exogenous water risks on their production and revenue, so long as it is profitable for them to do so (equation 1). Their reactions can be triggered for instance by extreme events (e.g. Mechler et al., 2010), or the deterioration of water quality with agriculture consequences or a gradual reduction in access to usable water (OECD, 2013). This is consistent with the observation that climate change adaptation generates private future benefits for farmers; therefore, they will have an incentive and should have a role to play to reduce the impact they may face in the future (e.g. Ignaciuk, 2015).

Table 4.1. Factors affecting the probability of farmers responding to future water risks

Key factors	Effect on the probability of reaction *θ*		Effect on the strength of responses
	Short-medium term	Long term	
Probability of risks	Positive effect	Positive effect	Positive effect
Price of products	Positive effect	Positive effect	n/a
Cost of water	Positive effect	Positive effect	n/a
Price of inputs	Ambiguous	Ambiguous	n/a
Information about risks or response	Positive effect	Positive but lower effect	More options possible
Capacity to respond	Minimum threshold to reach		Stepwise progression
Freedom to operate	Positive effect until a maximum		Stepwise progression
Risk aversion	Positive effect		Positive effect
Protection to risks	Possible negative effect		Negative effect
Discount rates	No effect	Negative effect	Negative effect (long term)
Projection uncertainties	No effect	Negative effect	Negative effect (long term)

Source: Authors own work, based in part on derivations from the model.

In detail, however, a range of factors will affect the probability of farmers' response (Table 4.1).

- A number of economic factors affect the probability of responses (Annex 4.A1), such as the effectiveness and possible foregone revenues associated with the reaction, but also how water risks affect the use of other (non-water) inputs, their costs, and the costs of water (i.e. water prices).
- Other constraints, non-accounted but possibly adaptable in the model, can alter the balance between action and inaction. In particular:
  - Insufficient information about the risks (φ unknown or ignored) or the response, especially when considering long-term risks that are not realised yet, can prevent reaction.
  - Farmers' capacity to respond (e.g. $\overline{\Pi_1} - \overline{\Pi_R}$ exceeding the budget of the farmer), and his or her lack of access to credit, can also prevent any response, so does their freedom to operate (policy environment, long-term, non-recoverable investments).

- Farmers' degree of risk-aversion may result in attributing more weight into risky situation (de facto inflating φ), which will increase the likelihood of response but may also result in costly "overreaction" (the choice of non-profitable alternative).
- On the other hand, the possible protection to risks (insurance) farmers have could partially reduce the likelihood of response (by lowering $(\overline{\Pi_1} - \Pi_1)$). This is consistent with findings of the climate change adaptation literature; e.g. Annan and Schlenker (2015) find that American farmers with insurance opt for more heat sensitive crops and do not have the incentive to engage into adaptation.
- In the long run, discount rates (not accounted for in the model) and real uncertainties on future water conditions (for which there is no robust information) and on technical solutions may also alter incentives to react. Even if the combination of the two may in fact reduce the effect on reaction.

Farmers will typically adapt to high water risks by changing practices or technologies, or by switching agriculture activities (Mendelsohn, 2016).[3] The choice of option will depend on the factors listed in Table 4.1, as well as the type of water risks and market conditions (relative profitability and availability of agriculture alternatives).[4] Assuming farmers face an observable reduction of available and usable water supply that will affect their production and revenue; they may increase their water use efficiency to produce at least as much with less water. When considering a water quality risk they may resort to change in practices, including, drainage, dilution or treatment. Alternatively, for both water quantity or quality risks, they may change agriculture activities towards more profitable or less risky sources of income.[5]

Each of the options involve trade-offs for individual farmers, that depend on the transaction costs and/or foregone benefits of the solution if no risk is actually realised (Annex 4.A1.2). In particular, changing practices will only occur if the costs of doing so are inferior to the risk avoidance benefits, and changing agriculture activity will only be profitable if the alternative provides more resilient and still viable revenue.

Figure 4.1. Market effects of water risk responses

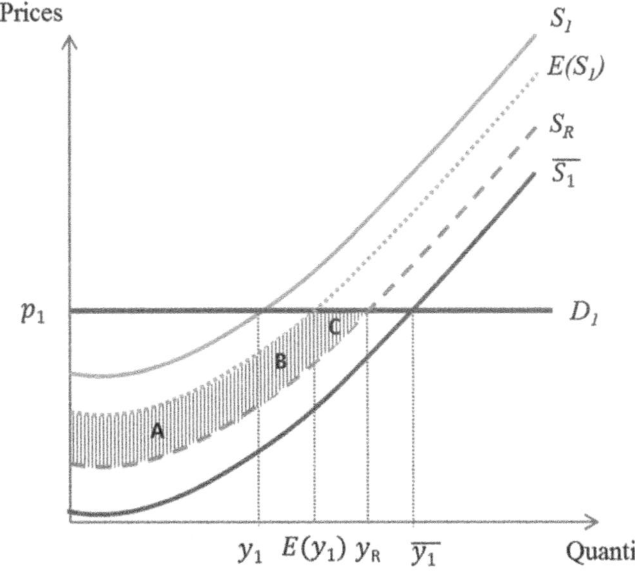

Note: S_1 is the supply with risk, $\overline{S_1}$ with no risk, $E(S_1)$ expected supply, and S_R supply with reaction. D_1 is the demand curve. This simplified illustration assumes a perfectly elastic demand and water risks are interpreted as a vertical shift in supply.

Source: Author's own work.

In each case, the solution will allow farmers to move away from maximum damages and improve their expected revenues, but it will not enable them to reach the profit level they would expect under no risk, because of the cost of response, as illustrated in Figure 4.1. Farmers will take action only if it allows them to obtain larger expected net revenue (with equilibrium quantity yR) than expected revenue with no reaction (quantity E(y1)). The gain of reacting is represented by the hatched area A+B+C in Figure 4.1, defined by the difference between the two supply curves E(S1) and SR under the demand curve.[6]

There are many examples of farmers' adaptation to exogenous water risks. The global development of groundwater irrigation in many regions was encouraged by decreasing precipitation and/or surface water variability (OECD, 2015a). In the United Kingdom, potato farmers responded to the 1975-76 drought by installing irrigation systems to better cope with rainwater shortfalls, and have not seen equally high production damages during subsequent droughts (Mechler et al., 2010). In the face of increasing dry conditions, many California growers have gradually shifted their activity towards higher value per water drop, and increased their water efficiency (Cooley et al., 2014). Farmers have also adapted to increased risk of salinity by changing their practice as observed in the Netherlands (van Duinen et al., 2015).

When faced with endogenous constraints, farmers face a slightly different set of incentive compatibility constraints. In this case, farmers' use of water and the impact they may have on water resources in the present affects their own future water risks. In the model, the probability of risks will depend on water uses and/or impact at present (Annex 4.A1). The incentive compatibility constraint (equation 1) remains essentially the same. Voluntary and anticipatory responses to water risks can be expected when the risks of not acting exceed the cost of taking action today. Responses to endogenous water risks also include changing practices, changing inputs and activities. Groundwater salinity in the irrigating region of the Californian Salinas Valley is an example whereby the risk of irreversible costs triggered actions of farmers in co-ordination with cities (OECD, 2015a). In Northern China, farmers have responded to groundwater shortage by taking the control of wells and pump assets, developing markets, adopting water savings and changing cropping patterns (Wang et al., 2007).

Yet the trigger to take action is not automatic, because of the possibility of free riding behaviour; the probability of future risks depends on the water use and/or impact of all farmers in the region at present. An individual farmer reducing water use or water impact in the short term, by switching to different activities or practices, will reduce water risks in the long term, thereby reducing the need (and probability θ) for any farmer to invest in future response. This situation depicts a Nash non-co-operative game; the outcome of a farmer's action depends on that of other farmers. Therefore, unless the farmer represents a significant share of water use in the hotspot region, or that s/he faces a sufficiently strong signals to account for future water risks (policy or investment return), s/he may not have sufficient incentive to act. Collective action by farmers can help overcome this individual incentive constraint, as observed in certain regions with water quality initiatives (OECD, 2013b).

Agro-food companies are more likely to respond when they strongly depend on products vulnerable to water risk or when they risk harm to their reputation

Agro-food companies in hotspot regions face different constraints. They purchase farmers' products, whose prices they may affect, and these products may only represent one input in their production chain. They rely on procured products from multiple farmers that may in some case be also located outside of the hotspot region. Therefore, farmers' water risks may not automatically result in significant costs for companies. Their profit will also depend on prices for finished products which could be unrelated to water risks. Relatively larger companies may also be subject to different types of impacts; beyond future production losses, not responding to endogenous water risks may affect their reputation, reducing their sales and the confidence of their financial investors.

As with farmers, risk probability and the price of output will increase their likelihood of response (higher stakes), but generally the response of food companies (and related organisations) will also vary largely by scope, market factors and dependency with local food supply (Annex 4.A1.3). Larger operating companies that rely on a broader market may not suffer significant losses from production risks in hotspot located farms; therefore, other things being equal, their threshold for response is higher. Smaller companies are more vulnerable to losses and will need to find ways to adapt more rapidly.

Under exogenous water risks, the type of responses will depend on the economic viability of the option and the scope of operations. Three options can be envisaged. First, companies may purchase products from farms outside of the hotspot region, if those alternative regions do not face an equally high risk; second, they may possibly change their agriculture inputs (or ingredients); and third, they may undertake stewardship programmes with farmers to ensure that they respond to risks (change practices or technologies). The choice of responses among these three options depends on transaction costs, such as transport costs for outside of the region engagement, cost of alternative ingredients, or alternative practices (Annex 4.A1.3). The response will also depend on the scope of operations; larger companies may be able to shift their purchases to other production units outside of the hotspot region in a temporary or permanent way, or engage with farmers. Smaller companies' existence may depend on supplies in the hotspot region.

Companies' incentive compatibility constraint will be altered when considering endogenous water risks caused by farmers and by their own operations. In this case, three factors change; first, farmers decision to react may matter, second, farmers contribution to overall risks, which could create uncertainties, and third, relatively large companies not reacting face reputation risks that may affect their sales (losses of revenues) or the confidence of investors on which they rely (e.g. affecting the companies' long term growth and investments). Reputation risks can be factored in as penalties (Annex 4.A1.3). Everything else being equal, because they face such reputation risks, larger companies are more likely to engage into a response than smaller ones.

More specifically, however, the question is whether larger companies will favour one approach or another. Moving away from water risk locations may have some reputation risks, as it avoids solving the problem and may displace the problem to other regions. Changing inputs could be helpful if the substitute effectively reduces the water risk factor and encourage farmers to produce alternative crops. Yet effectively engaging with farmers is probably the preferred option for consumers and investors. This option may evolve from stewardship initiatives to contracting and private standards, moving from voluntary options to mandatory requirements.[7] Market power may play a role in determining whether the approach will work; companies that have the largest agriculture market share can exert more power on their suppliers than smaller ones.

Smaller companies may approach the problem differently, as they are less likely to face reputation issues with investors or consumers. If such issues are not important, the problem they face will be similar to that of an exogenous risk, albeit facing more uncertainties in the future.

Examining the combined responses of farmers and companies: gaps, lack of co-ordination and strategic behaviour

Farmers and companies will respond to water risks differently depending on the type of farm and company and their respective sets of constraints. More resilient farmers will react to exogenous or endogenous risks to which a more vulnerable farmer may not be able. A large company may be more able to respond to different types of risks than a small company. If greater water risk will increase the average probability of response by all actors, it is difficult to predict the share of farms and companies that will effectively be able to tackle the risks proactively without knowing more about regional and market characteristics.

This heterogeneity of responses applies when considering the three regions of Northeast China, Northwest India and Southwest United States. Northeast China is populated by very small farms that may be vulnerable or semi-resilient to water risks. At the same time, the sector includes large agro-food companies, at least partially state-based,[8] who may contribute to risk mitigation especially if they are rated based on government objectives. In Northwest India, farmers can also be considered vulnerable or semi-resilient, while most food companies operate on a small to medium scale and may not be willing or able to respond to risks. In the Southwest United States, a large share of farms operate on a very large scale, with capacity to respond, and may act in conjunction with large co-operatives or companies to minimise the risks. Yet in all three regions, continued groundwater depletion also shows that endogenous water risks have not been satisfactorily addressed by farmers or companies.

In reality, uncertainties about future water risks may also matter, as these risks will be perceived differently among actors, possibly leading to co-ordination problems. In the above section, the behaviour of companies and farmers was analysed separately to understand the type of incentives they may respond to facing a possible but generally known risk. However, the risk is not only uncertain but it may also be perceived differently by different farmers and companies depending on their level of information and degree of risk aversion. For instance, farmers may believe that water risks are higher than companies do. Consequently, there may be asymmetric information and risk perception among farmers and among companies. In general, the actors with higher risk perceptions are more likely to react before other actors, assuming they share the same level of information. Better informed actors about risks will also act faster than others, assuming that they share the same degree of risk aversion.

The lack of information and possible information asymmetries may create co-ordination problems and lead to inefficient autonomous response to risks. If farmers get helped to mitigate risks, why do it on their own? If some companies act to reduce risk, why should others also contribute? Companies could do less than they would according to economic expectations, especially if farmers perceive the risks are higher than they perceive. Farmers may wait for companies to take any action and therefore under-adapt.

In the case of the three hotspot regions, the level of information about water risks does vary significantly (due to education, information access, capacity etc.). The three regions have already suffered long and intense drought spells, so awareness on water risks is high among all types of actors. However, information on the extent of the problem, its future projections, and the contribution of farmers to water challenges likely vary significantly from India to China and the United States. Large companies operating in the United States may also be more proactive than in the two other countries due to their capacity to provide flexible responses. Just like companies harvesting lettuce in California are able to change location three times a year to respond to climatic conditions, they may also be able to adapt to changing water supply and climatic conditions.

Well-defined government actions are needed to encourage and facilitate effective and co-ordinated responses

The role of government should be to incentivise farmers to engage in risk-reducing behaviours by increasing the benefits of acting or the cost of inaction, encouraging supply chain actors to also play a part of the role in risk mitigation activities. In particular, governments should address exogenous risks by increasing the probability of response by farmers (θ). This can be done, for instance, by providing more and better information on the cost of inaction and supporting effective response, but it may also involve regulatory instruments that will ensure that farmers take actions. For endogenous risks, governments have a role to play to redress information asymmetry to avoid co-ordination failures, but also to provide a signal for risk-enhancing farmers to take action, by employing regulatory or economic incentive measures (OECD, 2010b; 2014 and 2016f). For instance, increasing the price of water may encourage the uptake of practices or changes in crops under water risks (Annex 4.A1). Because risk-enhancing actions will generally continue, the government's engagement should have a long time horizon, with targets and monitoring, to ensure that efforts undertaken during a period are not wasted in a future period.

Still these actions will not be warranted in all contexts. Accounting for autonomous actions by farmers and companies, three main criteria appear necessary to any government intervention to address water risks in hotspot location.

- *Rationale*: Government's action will be warranted economically if it addresses market failures or market imperfection (OECD, 2014). Ignaciuk (2015) identifies seven criteria whereby public action is warranted for climate change adaptation ranging from knowledge development and dissemination to removing institutional barriers. Governments have a role to mitigate market failures associated with the risk of floods and droughts for agriculture (OECD, 2016f).
- *Additionality*: Government's engagement in hotspot water risks will only be effective if it adds significantly to existing policies towards agriculture risk management; it should drive either new or

more intense responses in hotspot locations, and avoid overlap with existing broader government efforts (such as information provision or investment in irrigation efficiency).

- *Targeting the right actors*: For government's action to enhance welfare, it should focus on the most likely impacted actors and account for autonomous reactions. Public action will only be effective if it aims at accounting for the capacity and incentives of farmers and companies to undertake autonomous actions, thereby identifying adaptation gaps.

The level of action is also an important consideration when planning for government's response to water risk hotspots. Adopting a hotspot approach will necessitate increased engagement on water risk mitigation in specific agriculture regions, either directly by national agencies or by devolving responsibilities to regional authorities. Local agriculture and water basin agencies and regional authorities (cities, district, counties, provinces or states) in charge of water and agriculture will need to be accounted for to ensure an effective response. National government agencies will therefore need to define their role in such system, be it as a facilitator or federator of responses, a complementary actor, an investor or funder, or by assuming a regulatory oversight role.

The analysis has shown that agriculture future water risk management is a responsibility that can be shared by farmers, companies and governments The next sections will consider a set of policy actions to mitigate water risks at and beyond hotspot locations.

4.3. A three-tier action plan for governments: From reducing hotspot risks to alleviating broader indirect impacts

A three tier action plan is presented in Figure 4.2. The scope of action is differentiated for the hotspot region, the market directly impacted (1st tier), and the indirectly impacted (2nd tier) countries—reflecting the three layers of impacts discussed in Chapter 3. Priority should be given to mitigating agriculture water risks in identified hotspot locations (1), not only for effectiveness and efficiency reasons, but because the more risk mitigation is conducted at this level, the lower impact there will be outside of the hotspot location. Recommended actions in sphere (2) and then (3) are conditional on the efforts undertaken at the hotspot location and may become secondary if hotspot level risks are well managed.

Figure 4.2. Proposed action plan for governments to address future water risk hotspots

Left hand side: Response by impact layer, Right hand side: scope for action for each type of country

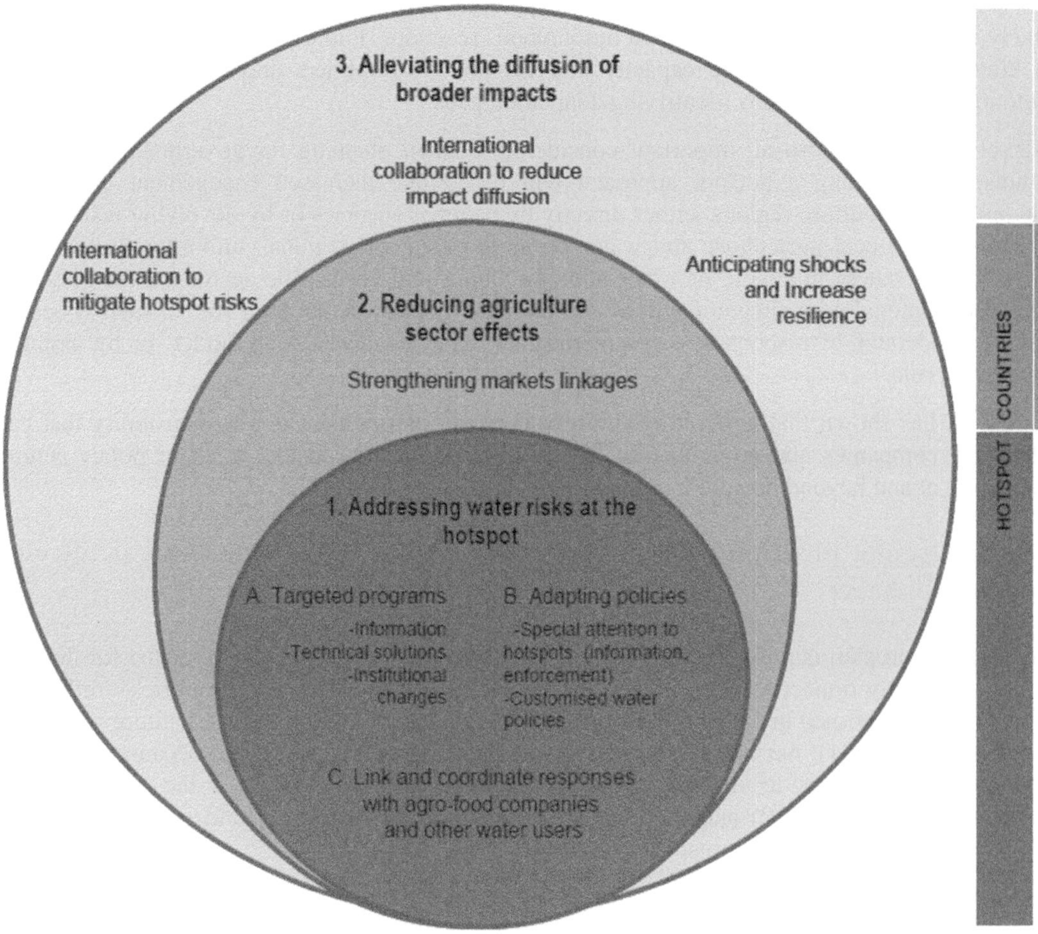

3. Alleviating the diffusion of broader impacts
International collaboration to reduce impact diffusion
International collaboration to mitigate hotspot risks
Anticipating shocks and Increase resilience
2. Reducing agriculture sector effects
Strengthening markets linkages
1. Addressing water risks at the hotspot
A. Targeted programs -Information -Technical solutions -Institutional changes
B. Adapting policies -Special attention to hotspots (information, enforcement) -Customised water policies
C. Link and coordinate responses with agro-food companies and other water users
HOTSPOT COUNTRIES
1st TIER COUNTRIES
2nd TIER COUNTRIES

Note: 1st tier country: market partner, 2nd tier country: other countries potentially affected.

Source: Author's own work.

The following sections describe these proposed actions, it first focuses on priority actions in the hotspot country and then moves to market and broader effects.

4.4. Reducing the risk at the source: Targeted programmes, adapted policies, and co-ordination with key stakeholders

The proposed action plan (Figure 4.2) incorporates three types of action for governments in hotspot locations. As for most regions, water risk hotspots require well-functioning generic water policies (OECD, 2016a). Beyond these, **two main policy approaches** can play a role in mitigating future agriculture water risk in hotspot regions. First governments can introduce additional targeted programmes or investments that add to pre-existing policies to better adjust to future risk; second, they can adapt existing national policies at the regional level to better respond to projected critical risks. Both approaches have merits, and may gain from being combined. As a complement to these approaches, **governments should co-ordinate their efforts** with relevant stakeholders. In particular, they should interact with and/or facilitate water stewardship efforts by agro-food sector companies. In addition, they should account for interaction of agriculture with other water users and the policies to which they are subject. The rest of this section will look at these three aspects (additional programmes, adapting policies, co-ordination with other actors) successively, encompassing different types of approaches based on the existing literature and illustration from existing programmes.

Targeted approaches to mitigate agriculture water risk in hotspot areas

Having identified future water risk locations, governments can decide to introduce additional programmes in hotspot locations to both increase effectiveness and reduce the cost of intervention. The level of action can vary by region, state or province, to a county or district, or even farms most at risk or most responsible for water risks, and could expand if successful in the hotspot area if funding allows.

Governments should engage into a combination of technical and institutional approaches to tackle water shortage risks. For agriculture water shortage risks, targeted initiatives should aim at reducing water consumption (demand side) or storing more water (supply side), and may include approaches geared toward technical or institutional solutions (solutions that require changes in the conditions under which farmers are operating) as shown in Table 4.2. These approaches may address both exogenous and endogenous water risks, and include investments, incentive programs, or new regulations.

Table 4.2. Targeted initiatives to tackle agriculture water shortage risks

	Demand side	Supply side
Technical solutions	-Establish water resource and water use monitoring systems -Targeted information programme to encourage context-specific water efficient farm practices -Cost-share programmes for efficient pressurised irrigation systems, precision agriculture sensors, or water conservation practices in the region -Support breeding for less water savvy and drought tolerant crops and livestock	-Conduct water supply vulnerability assessment -Targeted investment the development or maintenance of local dams, canals, and irrigation systems -Cost-share programmes into rainwater harvesting or infiltration ponds -Target investment in aquifer storage mechanisms (managed aquifer recharge, aquifer storage and recovery). -Invest in water reuse systems
Institutional solutions	-Regulating agriculture water use in the region or watershed -Setting up a charges for water used that varies with risks -Setting up water transfer or water market mechanisms -Measures favouring a shift away from water intensive crops	-Establishing regional groundwater banks

Source: OECD (2010, 2015a, 2016f); Cooley et al. (2016); Nam et al. (2015); and Ward (2016).

Governments should consider promoting technical solutions, via incentive, cost-share, R&D investment, or information instruments, as these have the advantage of enabling direct targeted actions on the source of the problem. On the demand side, they could invest in water monitoring and information systems, as these are constraints for farmers to act. They could also encourage farm management practices directed towards irrigation efficiency, which, if coupled with consumption reduction objectives, have a significant potential in addressing risks of water shortages in agriculture. For instance, Jägermeyr et al. (2016) use a projection model to simulate irrigation water productivity increases associated with different farm management practices globally under climate change scenarios. They find that irrigation efficiency improvements could lead to savings accounting up to 48% of non-productive water consumption. Governments could also promote precision agriculture and water conservation management practices as these have a potential to decrease water use (OECD, 2016d). In the Indian state of Haryana, Ambast et al. (2006) estimated that improvements in irrigation and fallowing practices for rainwater conservation and groundwater recharge would increase wheat productivity by 23% while potentially stabilising groundwater levels. Investing in regionally-focused applied breeding programmes towards less requiring or more drought resilient crop varieties and animal breeds may also contribute significantly to reduced water use, by favouring the switch of farmers to other relatively profitable opportunities why stabilising or lowering water consumption.

On the supply side, governments should focus their attention on addressing gaps, and facilitate or support investments in promising options for agriculture and water (Hanjra and Qureshi, 2010; Cooley et al., 2016). Investments in storage, if properly designed, can help sustain agriculture production under increasingly unstable precipitations (OECD, 2016d). Water reuse (or recycling) is increasingly seen as a necessity for operations to become less water thirsty in drought-prone regions (Hasler-Lewis, 2014). Irrigation development is also an effective means to reduce water shortage risks in drought-prone rainfed agriculture regions, particularly in developing countries (HLPE, 2015). In addition, in humid agriculture regions that expect to face future shortages, such as part of the United Kingdom, there may be some benefit to design supplemental irrigation schemes (Rey et al., 2014).

In recent years, targeted information efforts have been used to strengthen the resilience of farmers to climate change, including those related to future water shortage risks. The US Department of Agriculture has developed "climate hubs" (Box 4.1), or regional agencies covering specific agro-climatic regions, with the goal of assessing and responding to locally-specific agriculture or forest production vulnerability (USDA, 2016). A similar approach, focusing on irrigation vulnerability on the water supply side, has been conducted in Korea (Nam et al., 2015).

Box 4.1. Informing farmers of adaptation options: The USDA Climate Hubs

The US Department in Agriculture launched the Climate Hub regional programme in 2014, with the objective of "delivering science-based knowledge, practical information and programme support to farmers, ranchers, forest landowners, and resource managers to support climate-informed decision-making in light of the increased risks and vulnerabilities associated with a changing climate". Seven regional hubs were launched to cover the wide range of agro-climatic conditions spanning the United States. Within the Southwest Regional Climate Hub, California has been separated into a Sub Hub, to reflect the specificities of California's agriculture and climate challenges. Hubs rely on agents from regional or state offices of the USDA Agriculture Research Service or the USDA Forest Service, in collaboration with the USDA Natural Resources Conservation Service.

The Climate Hubs offer tools, management options and technical support to farmers, ranchers and landowners to help them adapt to climate change. The first activity undertaken in each hub or sub-hub is to use projected climate change impacts to map the vulnerability of different types of farms to climate change within the region. Hub or sub-hub agents then work with agronomic, livestock and forestry researchers and extension specialist to develop customised brochures and/or other information tools that recommend possible agronomic or technical responses to specific future climatic risks to different types of activity. The third step consists in disseminating these information brochures, and providing support to their application in local farm service agencies, with the support of land grand university extension specialists. In California, much of the vulnerability is related to the risk of water shortages; farmers have been proposed a menu of technical options to preserve water, store groundwater or become more efficient.

Source: USDA (2016) and personal conversation with USFS officer at the California Sub-hub.

Policies that provide financial support to technical approaches, even in a targeted manner, can have limitations if done in isolation of signals aiming at controlling total water use. In particular, support for irrigation efficiency has been found in a number of instances to be ineffective (OECD, 2015a; Ward, 2016). Shifting to more efficient irrigation in the absence of institutional controls on expansion of consumptive use can lead farmers to switch to crops that require more irrigation, reduce deficit irrigation or expand acreage, leading to an actual increase in water consumption.[9] This has been observed in northeastern Spain for instance, where irrigation modernisation has led to increased consumptive use (Lecina et al. 2009; Jimenez and Isidoro, 2012). Furthermore, increasing water efficiency can lead to reduce recharge of aquifers, reducing groundwater recharge, which further the problem of groundwater depletion in critical regions, like Northwest India (Ward, 2016) or the High Plains Aquifer (Pfeiffer and Lin, 2014). Dam or infrastructure investments also can result in conflict and lead to environment problems, and generally do not result in long term changes in the use of water.

Hotspottargeted institutional solutions may provide more lasting options for government to address water shortage (bottom row of Table 4.2). In particular, well-designed water transfers have the potential to help alleviate agriculture water risks in the United States Southwest (Cooley et al., 2016). In Australia, the multi-state water market around the Murray Darling Basin, regulated under an independent agency, provides incentive for farmers to be efficient and change agriculture activities based on the price signal they receive.

Since its introduction, farmers have been able to cope with largely fluctuating climatic conditions (Young, 2016). The system, however, required significant changes in resource allocation, involved prolonged efforts for years, and significant public investment. Encouraging nimble water transfers, adapted to local need and conditions, temporary and may also help in allowing conditional flexibility either for users in hotspot areas (nimble water transfers), or during severe weather events, potentially involving different set of actors (Colby, 2016). Measures that also discourage the use of relatively more water intensive agriculture activities may also have a role to play-reducing support for these crops or promoting the use of other activities may help in shifting the water consumption pattern in a hotspot region.

Targeted institutional solutions are especially relevant in the case of groundwater; successful groundwater management schemes generally involve collective action based management schemes that could effectively be used in water risk hotspot regions. The Nebraska Natural Resource District system is a successful example of management, whereby local actors decide of targeted approaches to sustainably manage groundwater use, to limit overdraft and its economic environmental consequences (Kassman, 2016). Groundwater banking and groundwater market initiatives often rely on local initiatives.

Naturally, addressing water shortage risks will be more effective with a combination of technical and institutional solutions. For instance, Saleth et al. (2016) analyse pathways to effective agriculture water demand management and show that institutions, infrastructures and technologies are necessary elements to an impact pathway to reduce agriculture's impact on water resources. Table 4.3 shows a suggested approach to address the problem of long-standing rapid depletion of groundwater in irrigated fields of Northwest India proposed in the literature.

Table 4.3. Proposed policy instruments to manage agriculture groundwater use in Northwest India

	Local	State and national	Regional / global
Information	-Generalisation of energy and/ or water metering systems	-Development of water databases (quantity, quality) -Aquifer mapping	
Technical Investment	-Water-saving technologies (e.g. drip irrigation, laser-levelling direct seeding) -Recycling of sewage water for irrigation -Groundwater recharge -Aquifer rehabilitation programmes		-Development of crop varieties resistant to salinity and drought
Economic	-Collective water resource management -Crop diversification -Alternative rural revenue	-Reform of the energy flat-rate system -Decoupling of subsidies with groundwater abstraction	
Regulatory		-New legal framework (Groundwater Bill) -Better application of the regulation in notified areas	
Multi-factor			-Mitigation of climate change

Sources: Badiani et al. (2012); CCAFS and CIMMYT (2014); Dasgupta (2016); Gleeson and Wada (2013); Hanasz (2014); OECD (2015a), Taylor et al. (2013).

Governments should consider multiple instruments to reduce flood risks

Governments can also resort to a combination of targeted technical and institutional solutions to manage flood water risks. There are increasingly sophisticated tools to identify and map flooding areas, incorporating future climate scenarios that can lead to targeted engineered responses at the territorial level (Poff et al., 2015). Governments can respond to floods by employing technical responses, such as by investing in infrastructure (dams, seawalls), emergency information and response systems (OECD, 2016f). At the production level, farmers can be encouraged to develop drainage system; they may also need to adapt their farming practices (Weather and Evans, 2009). Flood tolerant varieties of rice and other crops have been developed and successfully used in flood prone regions. Submergence-tolerant (SUB 1) rice varieties, developed by an international consortium or researchers using marker assisted selection, were commercially released in 2009-11, were rapidly adopted to reach an estimated 3.8 million farmers covering 2.5 million ha in 2012 in flood-prone regions of India, Nepal and Bangladesh (Ismail et al, 2013). Institutional solutions are less frequent. The development of integrated flood management encourages the use of payment for flooding protection (OECD, 2015c), in which farmers would get paid to provide fields for flooding. The United Kingdom is increasingly moving in this direction (OECD, 2015c; Weather and Evans, 2009).

Targeted efforts are well suited to mitigate water quality risks

Water quality risks are especially appropriate to the use of targeted initiatives around a hotspot region, because they tend to be locally heterogeneous and accumulating locally over time. The management of point source pollution is generally easier as it focuses on reducing pollution at the source. Diffuse pollution typically on the contrary spans over much more extensive areas, and will also differ significantly from one place to another, impacting water resources in a heterogeneous manner, leading to opportunities of acting on prioritised areas.

Governments can resort to a range of technical options to address endogenous or exogenous water quality risks in hotspot locations. Water quality measurements, investment in monitoring equipment, information programmes to support drainage and/or improved used of agriculture inputs, promoting precision farming, conservation agriculture and water treatment systems could be undertaken in areas most at risks. Given the complexity of water quality issues, the success of public efforts is not guaranteed and will rely on a careful design and monitoring system (Box 4.2). Governments can encourage and/or fund specialised breeding efforts; for instance the development of salt tolerant varieties of vegetables that can rely largely on seawater or brine aquifers has proven to be effective in the Netherlands and is now being envisaged in other salinity-prone regions (van Risselberghe, 2016). Additional technical solutions for the future may be under development; for instance, applied nanoscience present some potentially interesting solutions to control water quality impacts from agriculture (Dasgupta et al., 2016). Governments can also undertake larger investments in infrastructure to stop seawater level increases, slow salinity or other contamination of waterways to reduce future water risks.

Box 4.2. The challenge of reducing endogenous salinity risks in Australia

Australia is subject to different types of salinity. Primary salinity is naturally based affecting patches of land. Secondary salinity, results from water leaking to saline aquifers. The leaked water raise water tables, with saline water reaching the surface and then evaporating. Such phenomenon can result from the planting of shallow-rooted plants, that leave water leaking, or by excess water from irrigation.

Several programmes have been designed to address this challenge over the years. In 2000-07, the National Action Plan for Salinity and Water Quality spent a substantial amount (AUD 1.4 billion) on programmes spread into river catchment areas, to provide means to monitor, information, extension programmes, or grants to land manager to change their practices. Despite these substantial efforts, and acknowledging the complexity of the problem, several evaluations have found that the programme had not been able to effectively mitigate salinity notably because it failed to prioritise investments, set realistic targets, and support programmes towards long term goals (Pannell et al. 2014).

More recent programmes have focused on information and targeted investments. The National Water Quality Management Strategy proposes guidance on salinity trigger values and outlines ways to reduce salinity and adapt irrigation practices. The Australian Government's Water for the Future programme is promoting more efficient use of water on farms. The initiative "Caring for our Country" also will fund specific targeted programmes to address the risks in different contexts.

Source: Australian Government (2010); Pannell et al. (2010).

Targeted approaches have been employed to address diffuse nutrient pollution from farming zones in certain countries (OECD, 2017). These approaches typically include both technical and institutional components, coupling scientific assessment with policy response going almost to the farm level. In Scotland, for example, a targeted extension programme to reduce effluents prioritised pollution catchments, then launched data gathering exercises, awareness programmes for farmers, and one to one engagement with farmers to identify solutions and was ultimately able to significantly increase the uptake of best management practices by farmers (Aitken and Field, 2015). The Southland region of New Zealand developed a physiographic approach to water quality risks; incorporating results from different modelling exercises to target responses to farms that can generate the most groundwater nitrate contamination (see Annex 2.A1.3).

Water quality trading can be effective to tackle water risks locally and could potentially be extended to a wider number of areas facing endogenous water risk problems (Shortle, 2012; Rockefeller Foundation, 2015; OECD, 2017). While they require relatively elaborate institutional settings, such markets can prove effective in reducing diffuse pollution (OECD, 2012). Several schemes have been developed especially in North America, but also in New Zealand (e.g. OECD, 2015b).

Responding to multiple water risks in hotspot areas

Targeted water risk mitigation measures can also respond to multiple types of risks at once, albeit sometimes prioritising actions on one risk while addressing other. On the technical side, precision agriculture and irrigation efficient practices can for instance help reduce the use of inputs, e.g. via the use of fertigation (i.e. combining pressured irrigation system with the application of fertilisers). The Delta Programme in the Netherlands, while focusing on protection to seawater flooding, also works against intrusion of saline water into freshwater used in part for irrigation. Investments in multi-purpose dams, if well-designed and strategically placed, can provide protection from flooding; ensure a steady water supply, while also responding to a demand for energy and recreation (Naughton et al., 2017). More research may be needed in targeted solutions that can respond to multiple risks at once.

Given the range of options available and limited funding availability, governments should set up a prioritisation mechanism to identify which are the most cost-effective options. The design of the plan could be that used in other agro-environmental programmes (OECD, 2010a) or customised to water issues. For instance, Beher et al. (2016) propose a mechanism to prioritise programmes aiming at reducing marine run offs and apply it to sub-catchments adjacent to the Great Barrier Reef in Australia.

Enhancing existing agriculture risk management and water policies to address critical risks in hotspot regions

In parallel, governments should try to tailor broader policy levers in hotspot locations to respond to critical water risks. Water management often combines national and subnational (watershed) level policies to ensure that localised issues are treated in a differential manner in different hydrologic zones. For instance, watershed decentralised management is at the core of the EU Water Framework Directive. A hotspot approach would reinforce this approach by enabling even more flexibility to address localised critical water risks particularly for agriculture. This may comprise not only customised regional water policy responses but prioritisation of national agriculture or water policy programmes.

Table 4.4 reviews how past OECD policy recommendation to mitigate water risks in agriculture could be relevant and adapted to a hotspot management approach. Covered work includes analysis of agriculture risk management (OECD, 2011), as well as recommendations to address the challenges of adapting to climate change (OECD, 2014; Ignaciuk, 2015), to control groundwater intensive use (OECD, 2015a), and to mitigate the risk of floods and drought mitigation (OECD, 2016f).

Table 4.4. Adapting policy recommendations to address future agriculture water risk hotspots

Recommendation	Area*	Relevance and application to a hotspot management approach
Information support		
Build and maintain sufficient knowledge of groundwater resource and use;	GW	Special attention to hotspot areas
Establish information systems to support farmers, water managers and policy makers	WQ	Special attention to hotspot areas
Risk management		
Information, regulation and training to support market-based instruments for risk management	ARM	General-applies to hotspot areas
Managing catastrophic risks	ARM	General- applies to hotspot areas
Providing a clear framework for the allocation of risk-sharing and responsibilities, and facilitating the development of insurance products.	FD	Special attention to hotspot areas
Holistic approach to risk management	ARM	Special attention to hotspot areas
Water policy changes		
Manage surface and groundwater conjunctively where relevant; Favour instruments that directly target groundwater use; Prioritise demand–side approaches	GW	General- applies to hotspot areas
Take a holistic approach to agriculture pollution policies; Use a mix of policy instruments to address water pollution; Set realistic agriculture water quality targets and standards	WQ	General- applies to hotspot areas
Apply a "tripod" combination of regulatory, economic and collective management instruments to address intensive use of groundwater	GW	Special attention to hotspot areas
Take into account the Polluter-Pays-Principle to reduce agriculture water pollution.	WQ	Adaptable in hotspot areas by modulating instruments
Design policies to ensure long term sustainable water management; Improve crisis management of droughts and floods in agriculture by developing flexible instruments for water allocation, with economic instruments	CCWA FD	Adaptable in hotspot areas- with more flexible allocation systems and/or economic instruments
Enforcement		
Enhance the enforcement of regulatory measures	GW	Special attention to hotspot areas
Enforce compliance with existing water quality regulations and standards	WQ	Special attention to hotspot areas
Removing policy distortions		
Remove farm support that artificially increase risk exposure	CCWA	General- applies to hotspot areas
Remove disincentives for farmers' adaptive actions	AACC	General- applies to hotspot areas
Avoid non-water related price distorting policy measures, such as subsidies towards water intensive crops and energy	GW	General- applies to hotspot areas
Remove perverse support in agriculture to lower pressure on water systems.	WQ	General- applies to hotspot areas
Other recommendations		
Assess the cost effectiveness of different policy options to address water quality in agriculture.	WQ	General- applies to hotspot areas
Agronomic and supply-side measures as complement under high groundwater stress.	GW	Special attention to hotspot areas
Enable the development of private innovation and encourage public-private partnerships in agriculture technology R&D	AACC	Special attention to hotspot areas
Ensure infrastructure are climate-proof	AACC	Special attention to hotspot areas
Monitoring and evaluation of adaptation policies	AACC	Special attention to hotspot areas
Foster agriculture land as provider of floodplains and soil water retention services	FD	Special attention to hotspot areas

Note: * AACC: Adapting agriculture to climate change (Ignaciuk, 2015); ARM: agriculture risk management (OECD, 2011); CCWA: climate change water and agriculture (OECD, 2014); FD: mitigating floods and droughts (OECD, 2016f); GW: managing agricultural groundwater use (OECD, 2015a); WQ: controlling water quality (OECD, 2012).

Source: Author's own work, based on cited references.

Three categories of recommendations emerge, distinguishing actions based on their level of application and relevance for a hotspot approach.

- A first category regroups general policy recommendations that will need to be applied at the national level, with no differentiation in hotspot areas. This includes the removal of agriculture policies encouraging water misuse or the exposure of farmers to water risks, such as support for polluting inputs, or electricity subsidies for groundwater irrigation.
- Second, a large share of measures can be applied more forcefully or in priority in hotspot areas. This includes improving information systems, enforcing regulations in or strengthening groundwater policies. Each of these measures involves additional efforts by the national or local governments to mitigate water risks for farmers, without changing the direction of political actions.
- The third category of recommendations—pertaining to water allocation regimes and the use of economic instruments—could be specifically adapted to hotspot conditions.

Economic instruments could be adapted to tackle agriculture water risks in a hotspot area or even specifically targeted and designed for these areas. As discussed in Section 4.1, increasing the cost of water can encourage farmers to respond to shortage risks. Demand elasticities of irrigation largely vary across contexts, so water charges do not always result in any change in water consumption (OECD, 2010; 2015a). The question would then be how to introduce gradually stronger signals and how to ensure that economic instruments respond to future water risks. Water markets have the advantage of setting the price without requiring information about water conditions; they can respond to medium-term fluctuation and evolve over time (e.g. OECD, 2015d; Mendelsohn, 2016). The use of cap and trade, with a reduced cap over time, could help incentivise market actors to become water efficient faster. Charging water would require assessment and monitoring of the situation by an agency, but if politically feasible could drive efforts forward. Non-market quantitative restrictions can also be gradual based on the evolution of water risks. In all these cases, a clear target and transparent plan of action should be set over a long-term horizon.

Several national or state policies have been designed to enable locally differentiated responses. In California, the Sustainable Groundwater Management Act relies on semi-autonomous management in groundwater bodies. All groundwater bodies have to reach sustainability objectives by 2042, but they have the liberty to decide how to get there (see Box 4.3). In the case of water quality risks, Denmark has been introducing a targeted regulation for nitrate pollution into groundwater, imposing more strict oversight on farming zones that are more likely to lead to increased groundwater pollution (Højberg, 2016). The reform includes increased emphasis on zones that are more likely to impact aquifers (Ibid.).

Box 4.3. Adapting efforts to address groundwater overdraft in California

Groundwater depletion for agriculture irrigation is a major challenge in California. Groundwater is an important water source for California farmers, accounting for nearly 40% of irrigation withdrawals in average years (Maupin et al., 2014) and up to 60% in dry years. UCCHM (2014) estimates that over 60 km3 of groundwater have been lost in the Central Valley over the last half century. There are strong indications that groundwater overdraft is worsening: recent data show that the Sacramento and San Joaquin River Basins collectively lost nearly 31 km3 of groundwater between October 2003 and March 2010, or about 4.8 km^3 per year (Famiglietti et al., 2011). Overdraft is especially severe in the southern parts of the Central Valley, where groundwater levels have reached more than 33 meters below previous historic lows (CDWR, 2014). Overdraft has resulted in saltwater intrusion and other water quality impacts, significant land subsidence, lost water storage, and increased energy costs, among other adverse impacts.

Groundwater use has been unregulated in much of California. In response to worsening groundwater conditions, the state recently passed the Sustainable Groundwater Management Act of 2014 (SGMA). The act provides a framework for local authorities to manage groundwater supplies but allows for state intervention if necessary to protect groundwater resources. Specifically, it requires the formation of local agencies by mid-2017 and requires those agencies to adopt and implement local basin management plans by 2022. Additionally, it requires basins to achieve groundwater sustainability goals by 2040 in medium- and high-priority basins in critical overdraft and by 2042 in all other medium- and high-priority basins. While it remains to be seen the extent to which sustainability goals will actually be achieved, SGMA is an important step toward more rational and sustainable use of California's groundwater resources. In the long term, the intention of SGMA is that its implementation should increase the reliability of groundwater for agriculture and other users, leading in particular to a new long term –sustainable- equilibrium for irrigated agriculture.

Source: Cooley et al. (2016), citing CDWR (2014); Famiglietti et al. (2011); Maupin et al. (2014) and UCCHM (2014).

Governments should collaborate with and facilitate effective actions of the agro-food sector to reduce water risks[10]

As a complement to their responses targeted at farmers, governments will benefit from working with other supply chain actors (e.g. UNEP, 2016). A small number of medium to large agro-food companies have engaged in stewardship programmes focused on water risk mitigation, aiming at reducing the use of water in their production chain (water footprint) and the impact it may have on the local environment.[11]

These water stewardship initiatives widely vary in their scope, scale and design, and have known an evolution from technical fixes to multi-stakeholder partnerships (van der Heide et al., 2016). Involved companies start by conducting water risk assessment to select regions where water risks are most critical. Efforts to reduce the water risk and impacts tend to address risks occurring at the processing stage before those at the farm level. Stewardship programmes at the farm level employ a range of strategies; in the case of shortage risks, this includes reducing water demand and/or increasing water supply. Investments in technical solutions are often associated with incentives design to encourage farmers to change practices (e.g. via farm contracts or cost-share investments). If some programmes are still conducted only on suppliers' farms, leading companies have been progressively shifting their focus towards actions at the watershed level, realising the need to operate at that scale to avoid unwanted results (e.g. reduced consumption being compensated by additional consumption in other areas). Still, the issue of scale remains problematic in many contexts (WWF and IUCN, 2015).

This evolution has implied the need to engage into watershed-level partnerships with a range of local partners beyond primary suppliers (farmers), further implying a growing and multi-faceted involvement of local and national governments (Box 4.4). Governments have been asked to provide technical support and information sharing at a watershed scale, enable and facilitate discussions and partnerships with stakeholders, ensure equitable and sustainable outcomes, and warrant a good governance and regulation of water on which to base actions. Some of the leading agro-food companies on the agriculture-water front even consider that more government involvement is necessary to move forward on the reduction of water risks (Box 4.4).

Box 4.4. Managing water risks for agriculture in collaboration with the private sector: Key lessons from a workshop

The OECD, in partnership with the Dutch Ministry of Economic Affairs, organised a workshop to discuss and advance solutions to improve the management of agriculture water risks in the private sector on 9 November 2016. Participants included representatives of agro-food companies, associated non-profit organisations and OECD delegates. The following takeaway lessons were drawn from the discussion.

- There is a crucial need to **improve information on water risks**. Better data is needed particularly groundwater resources and water quality risks. Shared risk assessment indicators and solution methods are missing.
- **The success of managing agriculture water risks lies in public-private partnership at the local (catchment) level, involving multi-stakeholder engagement**. The local nature of water risk should be addressed with localised solutions, and the public good aspect of water demands collaboration. Because of these two specificities, responses to water risk ought to be holistic and contextualised.
- At the same time, **successful partnerships and collaborations are not simple to implement**. They require leaders who take initiative, convene relevant stakeholders, and facilitate the project. The public sector, in addition to setting regulation, has an important role in bringing the stakeholders together; however, leadership in projects does not have to be limited to the local public authorities. The public sector is also implicated in the scaling up of projects, as it can be helpful in shoring up the necessary financial support.
- **More general public-private partnerships are essential for successfully managing water risks**, acknowledging that the 'distance' between the private and public sector is not the only gap that needs to be bridged. Co-operation is also needed between: scientists and local stakeholders, different government bodies and levels, water experts, agriculture experts, financial experts, or within the private sector.
- **Moving beyond current efforts by the private sector will require government involvement**. Active agro-food companies are ever more energised to address water risk mitigation, but would need governments to help co-ordinate and complement their efforts to advance on sustainable water risk management. At the same time, government policies, including economic instruments could help incentivise response to water risks for the large share of companies that have yet to assess their water risks.

Source: OECD (2016c); the agenda, participants and presentations are available at: http://www.oecd.org/tad/events/workshop-managing-water-risks-for-agriculture.htm.

Governments should therefore find means to take advantage of developing water stewardship opportunities for effective water risk mitigation action. The specific role they may have will depend on the situation. Simply entering the conversation, sharing information on agriculture water risks, policy leverages, and scope for action, and providing input on the design of effective actions could already be valuable. Stronger engagement may lead to cost-effective solutions, that if sustained could lead to effective changes in farmers' practices. Sharing the burden of moving forward especially on the reduction of endogenous risks, if well designed, could help increase the cost-effectiveness of public action.

An example of effective model for partnership was developed by the 2030 Water Resource Group. This group, serving as an enabling entity supported by a consortium of companies, national and international funding agencies, and hosted by the International Finance Centre, convenes stakeholders from the public sector, private sector, and civil society to engage into action towards lowering identified supply-demand water gaps in regions facing critical water shortages. Its objective is to create partnerships at the country level by bringing in key stakeholders towards a common objective, identify acceptable solutions, funding sources and active partners to implement the proposed plan. The initiatives has known a few successes and been asked to engage into more regions in the future.

If local governments are typically called in first, national governments can also support the development and success of such partnerships. For instance, the US government has partnered with relevant stakeholders to share resources and efforts to respond to local resource challenges. Introduced as part of the U.S Agricultural Act of 2014, the USDA Regional Conservation Partnership Program (RCPP) supports the local implementation of USDA land and water resource conservation programmes at the watershed and/or regional landscape scale. Under this programme, the USDA selects projects of partnerships with farmers and a wide range of other land or water related stakeholders—Indian tribes, non-profit organisations, state and local governments, private industry, conservation districts, water districts, universities—to respond to identified resource challenge (USDA NRCS, 2017). USDA and its partners then leverage different sources of public and private funding to "assist producers with a broader set of land and water conservation activities designed to increase the restoration and sustainable use of soil, water, and wildlife and related natural resources across the landscape" (Schaible and Aillery, 2016). As of 2014-15, the RCPP funded USD 66 million of projects towards improving irrigated agriculture (Ibid.).

Governments should also co-ordinate agriculture policy responses with policies targeted at other types of water users

Water risks in hotspot areas will affect a large number of water users beyond actors in the agro-food sector. Policies aiming at reducing impact in the agriculture sector will benefit from co-ordinated efforts to tackle water risks for all sectors (Dinar, 2016; Hanjra and Qureshi, 2010). This may involve local and watershed and national co-ordination efforts around hotspot regional risks.

Governments drawing plans to address water risks especially in hotspot areas would benefit from accounting for the interactions among the key users of water (HLPE, 2015). Analysis of connections on the water-energy-food nexus, or the water-energy-environment nexus, have shown that some tangible interactions exist and can be relatively significant in certain regions (Grafton et al., 2016).

For instance, in Northwest India, the agriculture challenge posed by water risks cannot be separated from that of energy subsidies. The development of low price solar-powered pumps that would reduce electricity requirements and greenhouse gas emissions could also potentially increase depletion, raising the question of how government should adapt to these new technologies (Box 4.5). Undertaking assessment of current and projected water risks in each sector within the hotspot region will be necessary to reduce the likelihood of counteracting or overlapping projects and initiatives.

Box. 4.5. Solar pumps in Punjab: Opportunity or challenge?

Solar powering provides a green alternative energy for tubewells equipped with diesel or electric pumps, and could represent an important development lever in villages that are not connected to the grid. Nevertheless this new source of energy, providing free power with little time restriction, could encourage farmers to pump groundwater with excess, and accelerate withdrawal of aquifers (Shah, 2016; Closas and Rap, 2017). Until now, the installation of a drip-irrigation system was a prerequisite to receive subsidies for solar pumps (DSWC b; Pearson and Nagarajan, 2014). However, this condition might not be sufficient to change the irrigation practice of farmers, especially for rice and wheat, for which drip-irrigation systems are not yet used (Singh, 2015).

In 2000, Punjab began its first State programme for solar pump development. Since then, only 2 000 solar pumps were installed by farmers in the State, since subsidies were often not substantive enough to motivate the investment (Roy, 2015). Nevertheless, the State and Central Governments are pursuing their support programmes, and the cost of solar panels is falling, which suggests further development of solar pumps, threatening to increase the depletion of critical aquifers.

In order to solve this trilemma between clean energy, rural development and groundwater management, the International Water Management Institute (IWMI) initiated a pilot study in Gujarat named "Solar Power as a Remunerative Crop". The principle of SPaRC is to allow farmers to sell back the surplus of energy left after irrigation. Thus, the additional revenue creates an economic incentive to conserve water (CCAFS and CIMMYT, 2015). It should be underlined that the implementation of this system requires the connection of each solar installation to the grid, which can be an obstacle in poorly connected areas that need an alternative energy source the most.

Sources: CCAFS and CIMMYT (2015); Closas and Rap (2017); Pearson and Nagarajan (2014); Roy (2015); Shah (2016); Singh (2015).

Governments could also promote or facilitate rural-urban water co-operation to respond to a number of water risks (Civitelli and Gruère, 2016). A growing number of rural-urban co-operation mechanisms have linked farmers with cities to tackle water shortage, floods and water quality (OECD, 2015c). This includes successful examples of water transfers to address groundwater depletion in Japan, prominent programmes of payment for ecosystem services ensuring lower water pollution in catchment in France, Germany or the United States, and some emerging examples of flood plain policies in the United Kingdom (Ibid.).

4.5. Addressing the indirect effects of water risk hotspots

Addressing water risks at the production level, as discussed above, should limit sector concerns. Still, because these risks can take unexpected proportions and rely on other user actors, governments should also consider that supporting efforts at a higher economic level could help ensure that the remaining impacts of agriculture water risks are well contained and do not expand or intensify over time.

Strengthening agriculture and food markets for better resiliency

Well-functioning food markets play a fundamental role in diluting the damages, reducing price effects for consumers, and ensuring that regional effects remain limited. As shown in Chapter 3, a water shock in a limited number of regions can result in significant changes in trade flows, compensating gaps on the world market. In a non-functioning market, whereby products would not be able to flow easily from region to region with the associated price adjustments, agriculture production decrease would lead to higher price increases than in an open market, and depending on the price elasticity of demand, either increased cost for buyers or waiting lines.

Integrated markets allow products from other parts of a country or from the international market to compensate for production losses, depress local prices and alleviate shortages. Mechler et al. (2010) simulated the economic impact for agriculture of a drought (-12% of yields) first in the Guadiana Basin and second in broader Southern Spain. The first shock induced large production losses, but a limited reduction in added-value for local farmers, because the price of goods did not change much due to the presence of agriculture supplies in neighbouring regions. Neither did the shock affect local consumers significantly. In contrast, a simulated drought over all Southern Spain created larger damages and market effects. Farmers in Guadiana did not lose as much because of an increase of price, but the overall loss for consumers was much larger.

Similarly, international trade is a recognised key climate change and water risk adaptation mechanism especially in agriculture (HLPE, 2015; OECD, 2014). It allows populations suffering a decrease in production to access substitute products from other countries not facing the same climate and water risks, with reduced price increase consequences. It also allows water savings; for instance, de Fraiture et al. (2007) showed that irrigation consumption would have increased by 11%, had there not been any international trade in cereals in 1995. Jouanjean et al. (2016)'s global empirical analysis further shows that rainfall variation affects prices of agriculture commodities, and that trade flows are reallocated following extreme climatic and water-related events, acting as a compensation mechanism.

Certain types of markets can help mitigate the effect of water risks (OECD, 2014). Future markets have a particular role in risk management, in that they enable market players to lower the impact of climatic shocks over a season or more. Storage may also have a role so long as it does not affect markets. Well-developed markets for insurance that cover multiple heterogeneous regions, mutualising risk to better cover farmers and companies facing higher risks prepare for production shocks and ensure that they can recover from shocks.

Concerned governments should therefore strengthen and facilitate market linkages, which could involve three types of actions. Governments should enable markets, support transparency, and ensure that market transactions are effective and efficient. Impediments to well-functioning markets, such as high transaction costs or information asymmetries, will delay and or reduce market response, which could become problematic when facing water-related food security threats. Many countries may lack effective road, electricity and information infrastructure and the capacity to store and market goods. Second, Governments should also promote information exchange on markets to increase global preparedness to international food market shocks. The AMIS project aims to provide a monitoring tool for policy and market actors to prepare for price shocks and could be strengthened and expanded (HLPE, 2015). While these shocks may be due to change of market conditions, and/or policies, they may also be the result of climatic shocks in key production areas. Lastly, they should continue to encourage the removal of agriculture tariffs and non-tariff barriers on a multilateral basis. In particular, governments could aim at strengthening discipline on agriculture export restrictions to ensure that low-income water stressed countries can access food on the international market (HLPE, 2015).

Policy approaches to mitigate the risk of international chain reaction

As discussed in Chapter 3, critical water risks in one or more agriculture regions can diffuse indirectly to other countries. Acute water risks can lead to imbalanced land investments that can deprive a country of future water and food security. Agriculture water risks may contribute to food insecurity, social unrest, possible conflicts and/or migration outside of hotspot regions. These impacts can vary in scope and duration- ranging from a few countries to global consequences, and from a rapid shock to a sustained trend that last for years or more.

Three proposed government responses can be envisaged, following Chapter 3's characterisation of impact ripples. Governments in first or second tier countries can respond to risk spreading in three ways (Figure 4.3), first, they can try to help reduce the risk in the hotspot or first tier countries, second, they can aim at mitigate the impact from the two waves, or third they can focus on increasing the resilience of 2nd tier countries to these possible impacts. The first two ways will require international co-operation, the third will focus on national policies.

Figure 4.3. Three options can help respond to ripples of impact

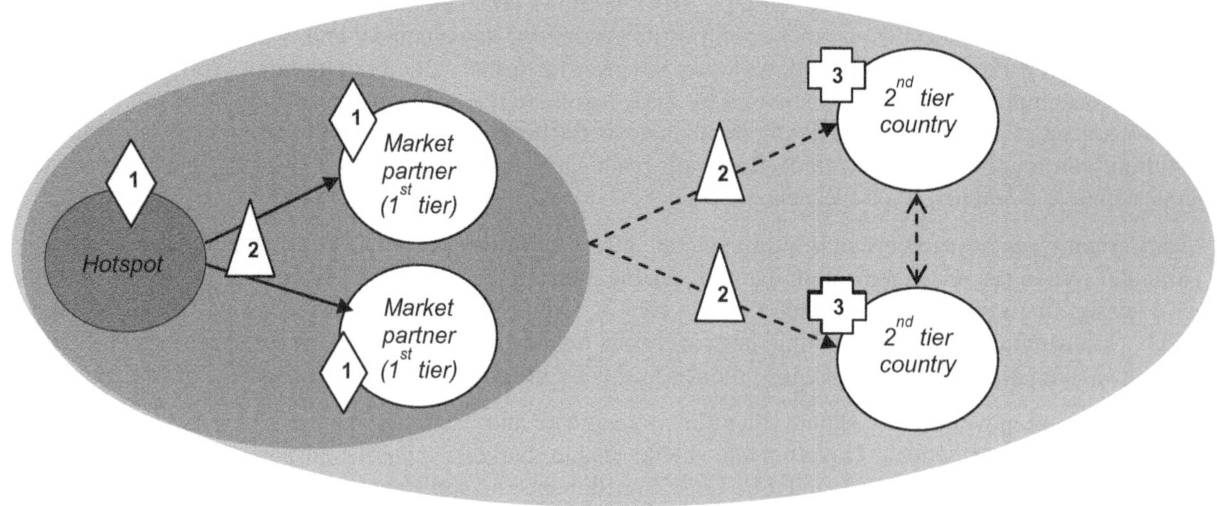

Notes: 1. Supporting risk reduction in the hotspot country and its market partners; 2. Mitigating impact diffusion; 3: increasing the resilience of 2nd tier countries.
Source: Author's own work.

International collaboration efforts to reduce agriculture water risks

Governments in the first and second tier countries can contribute to increasing resilience and lowering impact in hotspot locations bilaterally or via engaging more forcefully in international collaboration. They can participate in regional or international efforts by sharing information, expertise, co-operating on research and development, technology and services trade, or even contribute to the actual management of risks. Examples of these different linkages have emerged in the last few years.

Bilateral co-operation can take several forms depending on the characteristics of the involved countries. Government can support research and development exchanges on water risk monitoring, agriculture water efficiency improvements, lowering or treating water quality risks. There are multiple examples of such co-operation involving regions or countries facing high agriculture water risks. In 2015, the Californian Department of Food and Agriculture has exchanged with the Government of the Netherlands to facilitate exchange of expertise on climate-smart agriculture, including adaptation practices to water risks. The State of California has also exchanged with Israel on effective means to reduce agriculture water risks (Siegel, 2015). Water-depressed countries, such as China and the states in the West of the United States, have been exchanging with Australia's government and university experts to learn about the setting and functioning of the Murray Darling Basin's water markets. Within a development co-operation framework, the Technical Centre for Agricultural and Rural Cooperation (CTA), the joint co-operation institution of the EU and the African Caribbean and Pacific Group of States, includes a number of national programmes to reduce less water to produce more.[12] From 2007-16, the Japanese International Cooperation Agency (JICA) supported a project to increase water use efficiency in Tunisia (JICA, 2007).

Governments may also facilitate exchanges by concluding agreements between institutions, or by supporting international trade of technology and services. Israel's company expertise in highly efficient irrigation technologies has been promoted overseas as of 2011; Israel held 30% of the global market for advanced irrigation solutions (Ministry of Industry, Trade and Labor, 2011). Korea Rural Community Corporation, a state-based company affiliated with the Korean Ministry of Agriculture, Food and Rural Affairs, has been promoting new technology to better manage irrigation infrastructure and equipment (Ji-hye, 2014).

Governments may also engage in agriculture and water risk policy discussions at the international level. Several recent developments have shown an acceleration of interests of the international agriculture policy community on the challenges raised by high water risks. In 2015, the Committee on Food Security's High Level Panel of Experts on Food Security and Nutrition published a report on water for food security and nutrition, to respond to the fundamental question "How can the world ensure food and nutrition security given increasingly scarce water resources, especially in some regions, and the increasing competition for water uses?" (HLPE, 2015: 9). In 2016, under the Dutch EU presidency, the EU formed a taskforce dedicated to agriculture and water issues. In January 2017, under Germany's G20 presidency, agriculture ministers of the G20 countries signed a declaration and action plan entitled "Towards food and water security: Fostering sustainability, advancing innovation" which encompass commitments to address water scarcity, water quality and to reduce water risks (G20, 2017a and 2017b). In parallel to this event, 83 agriculture ministers the January 2017 participating to the Global Forum on Food and Agriculture (GFFA) adopted a Communiqué entitled "Agriculture and Water- Key to Feeding the World" outlining their intention to enhance farmers' water access, improve water quality, reduce water scarcity risks, and manage surplus water (GFFA, 2017).

International co-operation can also operate through support to global financing initiatives for climate change adaptation. Recent efforts facilitated by negotiations at the UN Framework Convention on Climate Change, including the development of the Green Fund, aim to support adaptation in developing countries, in which agriculture is often a key sector and one that is the most likely to be affected by climate change.

Multilateral trade negotiations on reducing barriers to environmental goods and services—as part of the Environmental Goods Agreement in particular, but also in regional trade agreements—may also have positive implications on risk mitigation. Eliminating barriers to trade of water monitoring devices, pressurised irrigation systems, and other related devices should facilitate their uptake in countries that can benefit most from those.

International collaborative efforts to mitigate the diffusion of primary and secondary impacts

International collaboration can also contribute to reduce the diffusion of impacts from water risks. As discussed in Chapter 3, agriculture water risks can increase investments in land markets that can create tensions and augment inequalities, and it can generate threats to food security with significant consequences on nearby countries, especially manifested via social and broader conflicts, political instability, and resulting migration.

Foreign land investments, while presenting risk mitigation potential, should be monitored on a case-by-case basis to ensure that land transactions do not result in detrimental water and food security outcomes for local populations. Land rich countries can benefit from such investments, and producing on land in a water abundant context can help reduce pressure on water resources in a highly water stressed country. But multiple examples have shown that it also can bear costs for non-represented stakeholders in the recipient countries.

Governments of contracted parties should ensure that the outcome of land investments remain positive in the long run. A number of conditions should be fulfilled to avoid unwanted outcomes from foreign land investments (as defined in a voluntary guideline, see FAO, 2012). In particular, there should be transparency in negotiations, deals should respect existing land and water rights, the benefits of the transaction should be shared with local communities, and the deals should be subject to a careful impact assessment to ensure their environmental sustainability (von Braun and Meinzen-Dick, 2009). Third party organisations, governments or international organisations could help support this goal by providing an external view.

Responding to food security and associated risks will solicit complex responses that overpass agriculture, and generally delve into development co-operation strategies. These include the role of emergency food aid, but also broader development co-operation programmes that support social safety nets, open and functioning markets, more stable political systems and good governance (Breisinger et al., 2015; OECD, 2013a). Effective co-operation is also needed to anticipate and manage migration flows (OECD, 2016e).

Anticipating and preparing for secondary impacts from agriculture water risks at the national level

The broad range of indirect consequences of agriculture water risks implies that proactive management of those risks in advance by governments of third-tier countries will be challenging. Just like hotspot regions, other countries should try and anticipate possible issues, be prepared to develop solutions, and/or to adjust policies accordingly.

Governments should monitor climate change projections and the evolution of agriculture risks in other countries. The hotspot identification exercise conducted in Chapter 2 shows that there is a growing body of evidence around a number future water risks globally. Uncertainties in assessment remain significant, but it is likely that more and better studies will be released in years to come.

Anticipating possible secondary impacts may also call upon the use of foresight exercises, building hypothetical scenarios and envisaging evolution of water crises, and possibly supporting these efforts with simulations. Prospective exercises can help anticipate complex scenarios for agriculture and explore options to strengthen resilience (OECD, 2016b).

Using methods that can adapt to changes of course—or follow adaptive pathways—could also help anticipate the risks. Haasnoot et al. (2013) propose a planning process that allows for changing situations. They consider likely scenarios and potential actions to mitigate the different water risks. Each of the action is then attributed a "sell-by-date", which indicate when a change in the course of action may be needed under each option. They then propose an efficient dynamic pathways to address the risks and work towards preferred pathways, each with initial action, threshold for change of course, and next step action. Haasnoot et al. (2013) apply the approach to salinity, water shortages and floods in the Rhine Delta, but similar approach could be used to tackle hypothetical indirect impacts from foreign water risks.

Countries anticipating migrating flows should try and anticipate and take action early to avoid future crises (OECD, 2016e). This encompasses anticipating future migratory flows and the needs for infrastructure and capacity; making pre-commitments to act if these flows are realised, and adapting existing policies to respond to crisis (Scarpetta, 2016, citing OECD, 2016e).

Governments in developing economies should also be encouraged to promote policies and investments to enhance food security, to increase the resilience of their farming populations (Hanjra and Qureshi, 2010). This includes increased investment in rural areas, policies to improve nutritional outcomes and to enhance agriculture productivity sustainably (OECD, 2013a). More broadly, these governments should pursue broad agriculture development agendas to strengthen the resilience of farmers, and reduce poverty (World Bank, 2008). The use of a territorial approach could help address food security while respecting the diversity of geographic, economic and governance systems (OECD/FAO/UNCDF, 2016).

4.6. Implications for the three hotspot regions

The capacity of farmers and companies to take action and the degree to which governments engage in water risk management, in particular regions, can provide an indicator of expected impacts in hotspot countries, market partners (1st tier) and potentially other countries (2nd tier) in the future. In particular, the impacts of the three agriculture production hotspot regions analysed in this report will likely vary significantly.

United States Southwest farmers are relatively resilient to water risks; many farmers have taken significant steps to adapt to drier conditions in recent years. Recent policy developments also suggest that California is taking proactive steps to manage water shortages (groundwater management and investments), even if their implementation will take time (full implementation is for 2040) and could therefore set the state in a low equilibrium.[13] Beyond drought emergencies, federal agencies recognised the importance of water issues in the region. The March 2016 Presidential Memorandum on building national capabilities for long-term drought resilience aimed to address drought impacts in the Western United States

(US White House, 2016). Other states in the Southwest are also increasing their efforts to reduce pressure on water. These efforts could reduce the impacts of water risks from this hotspot to other regions and countries through agriculture markets. State authorities could consider additional options, such as encouraging agriculture and urban water efficiency improvements, investing in water banks and recycled wastewater systems, and facilitate well-defined water transfers (Cooley et al., 2016). Still, under the current course, agriculture production will likely be altered, but continue to be a highly productive and large income generating sector in the region.

Farmers in Northeast China operate on very small land plots and remain relatively vulnerable to shocks. The Chinese government has launched important policy reforms to respond to water challenges, moving from a water supply augmentation policy to more demand control and quality emphasis. In 2011, it launched the three red line policies aiming at controlling water use, increasing water use efficiency, and reducing water pollution (Tan, 2014). A number of initiatives have been taken that focus specifically on agriculture water use, including the modernisation of irrigation systems (van Steenbergen et al., 2016) and the use of economic instruments (pricing and markets) (Liu and Speed, 2009; Wang, 2017). Agriculture policies are now moving towards the decoupling of agriculture support from inputs, which could improve water quality impacts from agriculture. These and other reforms demonstrate the Chinese government's demonstrated willingness to address water risks. The challenge will be to effectively implement these policies, especially in rural areas, and particularly in the Northeast region, to reinforce and upscale current efforts and promising initiatives to manage water imbalances and water quality problems. Therefore, whether water risks from this hotspot may impact other regions or countries will depend largely on the effective implementation of policies in the next decade.

Farmers in Northwest India also have limited capacity to adapt to water risks. India's federal government has launched large initiatives on water (River Ganga rejuvenation), including groundwater management proposed a groundwater model bill, and there are some signs of possible interest to move towards decoupling farmers' payments to become an income support (Gulati, 2016). At the same time, state policies in the Northwest supporting energy subsidies appear to remain in place, and the government promoted shift to solar pumps could accelerate groundwater depletion (Closas and Rap, 2017). Federal and state governments should aim at redirecting farmers' incentives, ensure that existing regulations are enforced, and consider engaging into new policy initiatives, possibly in partnership with the private sector to tackle the multi-faced challenges of the region. Under the current course, the region may take longer to come towards a lasting reduction of water risks for agriculture and in the meantime its water risks may still impact other regions and countries.

Similar qualitative assessment could be run for other agriculture regions and countries subject to high water risks, in particular those around the Mediterranean Sea, in Latin America, Sub-Saharan Africa or Southern Africa (Figure 2.2 in Chapter 2). Combining water and agriculture risks with an assessment of farm and companies' capacity to respond and a review of water and agriculture policy plans shall provide a first indication of whether the country could be facing agriculture damage, and potentially create further market and broader impacts in other countries. It will not be sufficient to ascertain expected impacts, but can be used to at least eliminate countries unlikely to face significant agriculture water risk-related damages, and to compare countries' potential damages with a reasonable confidence.

Notes

1. In particular, in the initial model agents a) have access to information and b) there is no market failure.

2. It is assumed that the expected losses are significant, as farmers are located in hotspot regions.

3. In the long run, farmers may also attempt to increase water storage (rainwater harvesting, infiltration ponds), or attempt to secure other water sources (via reuse, etc.) to reduce risks.

4. The discussion focuses on water scarcity and water quality risks, but the risks of flooding could also be considered, again via a change of practices or activities.

5. If these different options are not economically feasible, they may just use less water, produce less, fallow some or all of their land, or even sell their land and abandon farming in extreme cases.

6. This difference is positive and relatively significant. If however the probability of risks was lower, the difference would decline and could even become negative.

7. As such, their engagement can be assimilated to that of a "private regulator", to address the problem of non-co-operation across farmers.

8. Some of these companies may be subject to reputation risks when such risk is communicated to the public.

9. For any crop mix, efficient technologies increase the share of applied water to consumed water which reduces the costs of consumption, leading to the expansion of water consumption and irrigated acreage.

10. This subsection is largely based on the outcome of a workshop held by the OECD in November 2016. Information on the workshop is available at: http://oecd.org/tad/events/workshop-managing-water-risks-for-agriculture.htm.

11. Still, many other companies still have to consider even assessing their exposure to water risks. For instance Ceres (2015) reported that two third of the 37 large food sector companies they surveyed in the United States had not taken any action on water risks.

12. See http://spore.cta.int/en/dossiers/article/water-for-agriculture-a-focal-point-for-development.html.

13. For instance, as of September 2016, many farmers in the San Joaquin valley were continuing to dig deep wells and intensively withdraw groundwater noting that they would continue until SGMA regulations are put in place in 2020 (Sabalow, 2016).

References

Aitken, M. and S. Field (2015), "Practical Action on Farm to Reduce Rural Diffuse Pollution", In *International Fertiliser Society*, Proceeding 777. ISBN 978-0-85310-414-8 http://fertiliser-society.org/proceedings/uk/Prc777.HTM

Ambast, S.K., et al. (2006), "Management of declining groundwater in the Trans Indo-Gangetic Plain (India): some options", *Agricultural Water Management*, Vol. 82/3, pp. 279-296.

Annan, F. and W. Schlenker (2015), "Federal Crop Insurance and the Disincentive to Adapt to Extreme Heat", *American Economic Review*, Vol. 105/5, pp. 262-266.

Australian Government (2012), "Salinity and water quality", Department of Sustainability, Environment, Water, Population and Communities, Australian Government, Canberra.

Badiani, R. et al. (2012) "Development and the Environment: The Implications of Agricultural Electricity Subsidies in India", *Environment and Development,* Vol. 21/2, pp. 244–262.

Beher, J., et al. (2016), "Prioritising catchment management projects to improve marine water quality", *Environmental Sciences and Policy*, Vol. 59, pp. 35-43.

Breisinger, C. et al. (2015), "Conflicts and Food Insecurity: How Do We Break the Links?" In International Food Policy Research Institute, *2014-15 Global Food Policy Report*, IFPRI, Washington DC.

CCAFS (CGIAR Research Program on Climate Change, Agriculture and Food Security) and CIMMYT (International Maize and Wheat Improvement Centre) (2014), "Climate-Smart Villages in Haryana, India", Research Program on Climate Change, Agriculture and Food Security, New Delhi, India. https://cgspace.cgiar.org/rest/bitstreams/34314/retrieve

CDWR (2014), "Groundwater Basins with Potential Water Shortages and Gaps in Groundwater Monitoring", Sacramento, California. www.water.ca.gov/waterconditions/docs/Drought_Response-Groundwater_Basins_April30_Final_BC.pdf

Ceres (2015), "Feeding Ourselves Thirsty: How the Food Sector is Managing Global Water Risks", Ceres, Boston, MA https://www.ceres.org/resources/reports/feeding-ourselves-thirsty-how-food-sector-managing-global-water-risks

Civitelli, F., and G. Gruère (2016), "Policy options for promoting urban–rural cooperation in water management: a review", *International Journal of Water Resources Development*, DOI: 10.1080/07900627.2016.1230050. http://www.tandfonline.com/eprint/sJSA84V6zYxIfazFK3Bd/full

Closas, A., and E. Rap (2017), "Solar-based groundwater pumping for irrigation: Sustainability, policies and limitations", *Energy Policy*, Vol. 104, pp. 33-37.

Colby, B. G. (2016), "Water linkages beyond the farm gate: Implications for agriculture", *Federal Reserve Bank of Kansas City's Economic Review, Special issue, pp.47-74.* https://www.kansascityfed.org/~/media/files/publicat/econrev/econrevarchive/2016/si16colby.pdf

Cooley, H et al. (2016), "Water risk hotspots for agriculture: the case of the Southwest United States", OECD Food, Agriculture and Fisheries Paper No. 96, OECD Publishing, Paris. DOI : http://dx.doi.org/10.1787/5jlr3bx95v48-en

Cooley, H. et al. (2014), "Agricultural water conservation and efficiency potential in California", Issue Brief 14-05, Pacific Institute, Oakland, CA. http://pacinst.org/app/uploads/2014/06/ca-water-ag-efficiency.pdf

Dasgupta, K (2016), "In water-stressed Andhra, farmers sign pact to share ground water", *Hindustan Times*, 4 June 2016. www.hindustantimes.com/india-news/in-water-stressed-andhra-farmers-sign-pact-to-share-ground-water/story-tcv4mP5mXNBvE74vzqB7eN.html

Dasgupta, N., et al. (2016), "Nanoagriculture and water quality management", in Ranjan, S. et al (eds.) *Nanoscience in Food and Agriculture 1*, Sustainable Agriculture Reviews, 20, Springer International Publishing, Switzerland. DOI 10.1007/978-3-319-39303-2_1

De Fraiture, C. et al. (2007), « Looking ahead for 2050 : scenarios of alternative investment approaches", in Molden.D. (ed.), *Water for Food, Water for Life: A Comprehensive Assessment of Water Management in Agriculture,* International Water Management Institute, Colombo and Earthscan, London.

Dinar, A. (2016), "Dealing with Water Scarcity: Need for Economy-Wide Considerations and Institutions", *Choices*, Vol. 31/3. http://www.choicesmagazine.org/UserFiles/file/cmsarticle_517.pdf

Famiglietti, J.S. et al. (2011), "Satellites Measure Recent Rates of Groundwater Depletion in California's Central Valley", *Geophysical Research. Letters*, Vol. 38, L03403.

FAO (2012), "Voluntary Guidelines on the Responsible Governance of Tenure of Land, Fisheries and Forests in the Context of National Food Security", FAO Committee on Food Security, Rome. http://www.fao.org/docrep/016/i2801e/i2801e.pdf

Gleeson, T., and Y. Wada (2013), "Assessing regional groundwater stress formations using multiple data sources with the groundwater footprint", *Environmental Research Letters*, Vol. 8, No. 4, 044010.

GFFA (Global Forum for Food and Agriculture)(2017) "Agriculture and Water- Key to Feeding the World", GFFA Communiqué, 9th Berlin Agriculture Ministers' Conference 2017, 21 January 2017, Berlin. http://www.gffa-berlin.de/wp-content/uploads/2017/01/GFFA-Kommunique_2017_EN.pdf

Grafton, Q. et al. (2016), "Responding to Global Challenges in Food, Energy, Environment and Water: Risks and Options Assessment for Decision-Making", *Asia and the Pacific Policy Studies*, Vol. 3/2, pp. 275-299.

G20 (2017a), "G20 Agriculture Ministers' Action Plan 2017- Towards food and water security: Fostering sustainability, advancing innovation", G20 German Presidency, Berlin, January 22 2017. http://www.bmel.de/SharedDocs/Downloads/EN/Agriculture/GlobalFoodSituation/G20_Action_Plan2017_EN.pdf?__blob=publicationFile

G20 (2017b), "G20 Agriculture Ministers' Declaration 2017-Towards food and water security: Fostering sustainability, advancing innovation", G20 German Presidency, Berlin, January 22 2017. http://www.bmel.de/SharedDocs/Downloads/EN/Agriculture/GlobalFoodSituation/G20_Declaration2017_EN.pdf;jsessionid=48AF37D05D7C0366B55297B7217593BE.2_cid376?__blob=publicationFile

Gulati, A. (2016), "Plug those subsidy leakages", *OECD Observer*, Vol. 306, p.16.

Haasnoot, M., et al. (2013), "Dynamic adaptive policy pathways: A method for crafting robust decisions for a deeply uncertain world", *Global Environmental Change*, Vol 23/2, pp. 485–498. DOI:10.1016/j.gloenvcha.2012.12.006.

Hanasz, P. (2014), "The problem with problems of water scarcity in South Asia", GWF Discussion Paper 1408, Global Water Forum, Canberra.

Hanjra, M.E. and M.E. Qureshi (2010), "Global water crisis and future food security in an era of climate change", *Food Policy,* Vol. 35, pp. 365-377.

Hasler-Lewis, C. (2014) "On the Future of Agriculture: New Technology, Water Recycling and Other Advances Should Reduce Our Farms' Thirst", *Wall Street Journal*, July 7 2014, New York.

HLPE (High Level Panel of Experts on Food Security and Nutrition) (2015), "Water for food security and nutrition", A report by the High Level Panel of Experts on Food Security and Nutrition of the Committee on World Food Security, Rome.

Højberg, A. L. "Utilizing natural nitrogen reduction in national regulation", Presentation at the International Conference Towards sustainable groundwater in agriculture, Burlingame CA, June 2016. https://www.youtube.com/watch?v=rbn-As5M68c

Ignaciuk, A. (2015), "Adapting Agriculture to Climate Change: A Role for Public Policies", OECD Food, Agriculture and Fisheries Papers, No. 85, OECD Publishing, Paris. http://dx.doi.org/10.1787/5js08hwvfnr4-en

Ismail, A.M. et al. (2013), "The contribution of submergence-tolerant (Sub1) rice varieties to food security in flood-prone rainfed lowland areas in Asia", *Fields Crops Research*, Vol. 152, pp. 83-93.

Jägermeyr, J. et al. (2016), "Integrated crop water management might sustainably halve the global food gap", *Environmental Research Letters*, Vol 11., 025002.

JICA (Japan International Cooperation Agency) (2007), "JBIC Signs ODA Loan Agreements with Tunisia – press release", JICA, Tokyo. http://www.jica.go.jp/english/news/jbic_archive/autocontents/english/news/2007/000071/reference.html

Ji-hye, S. (2014), "Korea promotes rural development model", *The Korean Herald*, 19 September. http://www.koreaherald.com/view.php?ud=20140919000724

Jimenez, M. and D. Isidoro (2012), "Efectos de la modernización de la comunidad de regantes de Almudevar (Huesca) sobre el cultivo del maíz", *Tierras de Castilla y León*, Vol. 193, pp. 102-109.

Jouanjean, M.-A. et al. (2016), "Price, water and trade in agriculture : an alternative to the rain dance", PRISE working paper, ODI, London.

Kassman, K.G. (2016), "Long-Term Trajectories: Crop Yields, farm land, and irrigated agriculture", *Federal Reserve Bank of Kansas City's Economic Review*, Special issue, pp.21-46. https://www.kansascityfed.org/~/media/files/publicat/econrev/econrevarchive/2016/si16cassman.pdf

Lecina, S. et al. (2009), "Efecto de la modernización de regadíos sobre la cantidad y la calidad de las aguas: la cuenca del Ebro como caso de estudio", Monografías INIA, Serie agrícola, No. 26, INIA, Madrid.

Liu, B., and R. Speed (2009), "Water Resources Management in the People's Republic of China", *International Journal of Water Resources Development*, Vol.25, pp.193-208

Maupin, M. et al. (2014),"Estimated Use of Water in the United States in 2010", US Geological Survey, Circular 1405, USGS, Washington D.C. http://pubs.usgs.gov/circ/1405/pdf/circ1405.pdf

Mechler, R. et al. (2010), "Modelling economic impacts and adaptation to extreme events: Insights from European case studies", *Mitigation and Adaptation Strategies for Global Change*, Vol. 15, pp. 737-762.

Mendelsohn, R. (2016), "Adaptation, Climate Change, Agriculture, and Water", *Choices*, Vol. 31/3. http://www.choicesmagazine.org/UserFiles/file/cmsarticle_520.pdf

Ministry of Industry, Trade and Labor (2011), "Israel's water sector", Israel NewTech, National energy and water program, Investment promotion center presentation, Tel Aviv. http://israelnewtech.gov.il/DocLib/%D7%97%D7%95%D7%9E%D7%A8%D7%99%D7%9D%20%D7%A9%D7%99%D7%95%D7%95%D7%A7%D7%99%D7%99%D7%9D/Israel%20NewTech%20-%20Water%20Sector.pdf

Nam, W. H. et al. (2015), "Irrigation vulnerability assessment of agricultural water supply risk for adaptive management of climate change in South Korea", *Agricultural Water Management*, Vol. 152, pp. 173-187.

Naughton, M., et al. (2017), "Managing multi-purpose water infrastructure: A review of international experience", OECD Environment Working Papers, No. 115, OECD Publishing, Paris. DOI: http://dx.doi.org/10.1787/bbb40768-en

OECD (2017), *Diffuse Pollution, Degraded Waters: Emerging Policy Solutions,* OECD Publishing, Paris,http://dx.doi.org/10.1787/9789264269064-en.

OECD (2016a), "Agriculture and water", Agriculture Policy Note, OECD, Paris, http://oecd.org/tad/policynotes/agriculture-water.pdf.

OECD (2016b), *Alternative Futures for Global Food and Agriculture*, OECD Publishing, Paris, http://dx.doi.org/10.1787/9789264247826-en.

OECD (2016c), "Managing water risks for agriculture: A discussion with the private sector", Summary of the workshop, OECD, Paris, http://www.oecd.org/tad/events/WorkshopSummaryJan2017.pdf.

OECD (2016d), *Farm Management Practices to Foster Green Growth*, OECD Publishing, Paris, http://dx.doi.org/10.1787/9789264238657-en.

OECD (2016e), *International Migration Outlook 2016*, OECD Publishing, Paris, http://dx.doi.org/10.1787/migr_outlook-2016-en.

OECD (2016f), *Mitigating Droughts and Floods in Agriculture: Policy Lessons and Approaches*, OECD Studies on Water, OECD Publishing, Paris, http://dx.doi.org/10.1787/9789264246744-en.

OECD (2015a), *Drying wells, rising stakes: Towards Sustainable Agricultural Groundwater Use*, OECD Studies on Water, OECD Publishing, Paris, http://dx.doi/10.1787/978926423870-en.

OECD (2015b), "The Lake Taupo Nitrogen Market in New Zealand: A Review for Policy Makers", OECD Environment Policy Papers, No. 4, OECD Publishing, Paris, http://dx.doi.org/10.1787/5jrtgl13p9mr-en.

OECD (2015c), *Water and Cities: Ensuring Sustainable Futures*, OECD Studies on Water, OECD Publishing, Paris, http://dx.doi.org/10.1787/9789264230149-en.

OECD (2015d), *Water Resources Allocation: Sharing Risks and Opportunities,* OECD Publishing, Paris, http://dx.doi.org/10.1787/9789264229631-en.

OECD (2014), *Climate Change, Water and Agriculture: Towards Resilient Systems*, OECD Studies on Water, OECD Publishing, Paris, http://dx.doi/10.1787/9789264209138-en.

OECD (2013a), *Global Food Security: Challenges for the Food and Agricultural System*, OECD Publishing, Paris, http://dx.doi.org/10.1787/9789264195363-en.

OECD (2013b), *Providing Agri-environmental Public Goods through Collective Action,* OECD Publishing, Paris, http://dx.doi.org/10.1787/9789264197213-en.

OECD (2012), *Water Quality and Agriculture: Meeting the Policy Challenge*, OECD Studies on Water, OECD Publishing, Paris, http://dx.doi.org/10.1787/9789264168060-en.

OECD (2011), "Risk management in agriculture: what role for governments", OECD, Paris. https://oecd.org/agriculture/agricultural-policies/49003833.pdf.

OECD (2010a), *Guidelines for Cost-effective Agri-environmental Policy Measures*, OECD Publishing, Paris, http://dx.doi.org/10.1787/9789264086845-en.

OECD (2010b), *Sustainable Management of Water Resources in Agriculture*, OECD Studies on Water, OECD Publishing, Paris, http://dx.doi.org/10.1787/9789264083578-en.

OECD/FAO/UNCDF (2016), *Adopting a Territorial Approach to Food Security and Nutrition Policy*, OECD Publishing, Paris, http://dx.doi.org/10.1787/9789264257108-en.

Pannell, D.J. and A.M. Roberts (2010), "Australia's National Action Plan for Salinity and Water Quality: a retrospective assessment", *Australian Journal of Agricultural and Resource Economics*, Vol. 54, pp. 437-456.

Pearson, N. and G. Nagarajan (2014), "Solar water pumps wean farmers from India's archaic grid", Bloomberg, www.bloomberg.com/news/articles/2014-02-07/solar-water-pumps-wean-farmers-from-india-s-archaic-grid.

Pfeiffer, L and C.Y. Lin (2014), "Does efficient irrigation technology lead to reduced groundwater extraction?", *Journal of Environmental Economics and Management*, Vol. 67/2, pp. 189-208.

Poff, N. L. et al. (2015), "Sustainable water management under future uncertainty with eco-engineering decision scaling", *Nature Climate Change*, http://dx.doi.org/10.1038/NCLIMATE2765.

Rey, D. et al. (2014), "Modelling and mapping the economic value of supplemental irrigation in a humid climate*", Agriculture Water Management*, vol. 173, pp. 13-22.

Rockefeller Foundation (2015), "Incentive-based instruments for water management: A synthesis review", Report for the Evaluation Office, written in collaboration with the Pacific Institute and the Foundation Center, The Rockefeller Foundation, New York, https://assets.rockefellerfoundation.org/app/uploads/20160208145427/issuelab_23697.pdf.

Roy, V.C. (2015), "Punjab farmers show little interest in solar water pumps", *Buisness Standard*, 6 April, www.business-standard.com/article/economy-policy/punjab-farmers-show-little-interest-in-solar-water-pumps-115040601135_1.html

Sabalow, R. (2016), "Farmers say, 'No apologies,' as well drilling hits record levels in San Joaquin Valley", The Sacramento Bee, 25 September, http://www.sacbee.com/news/state/california/water-and-drought/article103987631.html#storylink=cpy

Saleth, R.M. et al. (2016), "Role of institutions, infrastructures, and technologies in meeting global agricultural water challenges", *Choices,* Vol. 31/3, http://www.choicesmagazine.org/UserFiles/file/cmsarticle_521.pdf.

Scarpetta, S. (2016), " OECD countries need to address the migration backlash", OECD Insights, 19 September, http://oecdinsights.org/2016/09/19/oecd-countries-need-to-address-the-migration-backlash/.

Shah, T. (2016), "Political economy of energy-food-groundwater nexus: Toward drought and climate resilience", Presentation at the South Asia Groundwater Forum, Jaipur, 1 June 2016.

Schaible, G.D. and M. Aillery (2016),"Challenges for US irrigated agriculture in the face of emerging demands and climate change", in Ziolkowska, J.R. and J. M. Peterson (eds.), *Competition for Water Resources: Experiences and Management Approaches in the U.S. and Europe*, Elsevier Publishing, Amsterdam.

Shortle, J. (2010), "Water quality trading in agriculture", Background report, OECD, Paris, https://www.oecd.org/tad/sustainable-agriculture/49849817.pdf

Siegel, S. (2015), *Let there be water: Israel's solution for a water-starved world,* St Martin's Press, New York.

Singh, S. (2015), "Jain see s drip irrigation as social capitalism", Plastic News, 5 November 2015, www.plasticsnews.com/article/20151105/NEWS/151109916/jain-sees-drip-irrigation-as-social-capitalism.

Tan, D. (2014), "The State of China's Agriculture", China Water Risk, Hong Kong, China, http://chinawaterrisk.org/resources/analysis-reviews/the-state-of-chinas-agriculture/

Taylor, G. et al. (2013), "Groundwater and climate change", *Nature Climate Change*, Vol 3/4, pp. 322-329.

UCCHM (University of California Center for Hydrologic Modeling) (2014), "Water Storage Changes in California's Sacramento and San Joaquin River Basins from GRACE: Preliminary Updated Results for 2003-2013", UCCHM Water Advisory #1, University of California Irvine, CA.

UNEP (United Nations Environment Program) (2016), "Food Systems and Natural Resources", a report of the Working Group on Food Systems of the International Resource Panel, written by Westhoek, H. et al.,. UNEP, Nairobi. http://www.unep.org/resourcepanel/knowledgeresources/assessmentareasreports/food

USDA (U.S. Department of Agriculture) (2016), USDA Regional Climate Hubs Factsheet, USDA, Washington DC. hhttp://www.climatehubs.oce.usda.gov/sites/default/files/USDA Regional Climate Hubs Factsheet 2016.pdf

USDA NRCS (USDA Natural Resources Conservation Services) (2017), "Regional Conservation Partnership Program", USDA, Washington DC. https://www.nrcs.usda.gov/wps/portal/nrcs/main/national/programs/farmbill/rcpp/ (accessed January 2017)

US White House (2016), "Presidential Memorandum: Building National Capabilities for Long-Term Drought Resilience", Office of the Press Secretary, The White House, Washington DC. https://www.whitehouse.gov/the-press-office/2016/03/21/presidential-memorandum-building-national-capabilities-long-term-drought

van der Heide, M., et al. (2016), "Approaches to water risk management: illustrative examples of agriculture and food companies", Discussion note for the OECD/ Netherlands workshop on "Managing water risks for agriculture: A discussion with the private sector", November 2016. http://www.oecd.org/tad/events/Approaches%20to%20water%20risk%20management_OECD2016_2.pdf

van Duinen, R., et al. (2015), "Coping with drought risk: empirical analysis of farmers' drought adaptation in the south-west Netherlands", *Regional Environmental Change*, Vol. 15/ 6, pp. 1081-1093.

van Risselberghe, M. (2016), "Facilitating saline agriculture in order to solve food security, solution for climate change and solving poverty", Presentation of the Salt farm Texel and the Salt farm foundation, given at the OECD workshop on "Managing water risks for agriculture: a discussion with the private sector", in Paris, November 9 2016. http://www.oecd.org/tad/events/Session%202%202.Salt%20Farm%20Texel%20-%20OECD%20Paris.pdf

van Steenbergen, F. et al. (2016), "Agricultural production and groundwater conservation: Examples of good practices in Shanxi province, People's Republic of China", Asian Development Bank, Mandaluyong City, Philippines.

von Braun, J. and R.S. Meinzen-Dick (2009), "'Land grabbing' by foreign investors in developing countries: risks and opportunities", IFPRI Policy Brief 9, International Food Policy Research Institute, Washington DC. http://ebrary.ifpri.org/utils/getfile/collection/p15738coll2/id/14853/filename/14854.pdf

Wang, J. (2017), "Agricultural groundwater use in China: challenges, solutions and outlook", in Döll, P., et al. "Agriculture and Groundwater: Feeding Billions from the Ground Up", Presentation at the Global Forum for Food and Agriculture (GFFA), Berlin, January 20, 2017. https://www.slideshare.net/ifpri/agriculture-and-groundwater-feeding-billions-from-the-ground-up

Wang, J. et al. (2007), "Agriculture and groundwater development in northern China: trends, institutional responses, and policy options", *Water Policy*, Vol. 9/ Supplement 1, pp. 61-74.

Ward, F. (2016), "Policy challenges facing agricultural water use: An international look", *Water Economics and Policy*, Vol. 2/3. DOI: 10.1142/S2382624X1671003X

Weather, H. and E.Evans (2009), "Land use, water management and future flood risk", *Land Use Policy*, Vol. 26S, pp. S251-269.

World Bank (2008), "Agriculture for Development", World Development Report 2008, World Bank, Washington DC. https://openknowledge.worldbank.org/handle/10986/5990

WWF and IUCN (World Wide Fund for Nature and International Union for the Conservation of Nature) (2015), Water Stewardship in Agriculture, Brief, WWF International and IUCN, Geneva, Switzerland. https://d2ouvy59p0dg6k.cloudfront.net/downloads/waterstewardshipagri_lowres.pdf

Young, M. (2016), "Water Allocation in the West: Challenges and opportunities", *Federal Reserve Bank of Kansas City's Economic Review*, Special issue, pp. 101-128. https://www.kansascityfed.org/~/media/files/publicat/econrev/econrevarchive/2016/si16young.pdf

Annex 4.A1. Model derivation

4.A1.1. Derivation of equation (1)

$$E(\Pi/R) \geq E(\Pi/NR) \Leftrightarrow \varphi.\Pi_R + (1-\varphi).\overline{\Pi_R} \geq \varphi.\Pi_1 + (1-\varphi).\overline{\Pi_1}$$
$$\Leftrightarrow \varphi.(\Pi_R - \overline{\Pi_R} + \overline{\Pi_1} - \Pi_1) \geq \overline{\Pi_1} - \overline{\Pi_R}$$
$$\Leftrightarrow \varphi.[(\Pi_R - \overline{\Pi_R}) - (\Pi_1 - \overline{\Pi_1})] \geq \overline{\Pi_1} - \overline{\Pi_R}$$

4.A1.2. Modelling the farmer's behaviour

Assume that an individual price taking farmer's profit can be expressed as $\Pi = py - p_w w - p_x.\boldsymbol{X}$, where p is the farm gate price, y farm's output, p_w the expected cost of water (price or otherwise) [1] and w water used, and $p_x.\boldsymbol{X}$ representing other input costs (price times a vector of quantity). Further assume the following log linear production function: $y = \alpha w^{k_w}.\boldsymbol{X}^{k_x}$ with α a positive coefficient for any activity and $k_w$and k_x are production coefficients.

The profit function of the farmer without reaction under risks and no risks at t_1 are, respectively:

$$\Pi_1 = p_1\alpha_1 w^{k_w}.\boldsymbol{X}^{k_x} - p_w w - p_x.\boldsymbol{X} \quad (3)$$
$$\overline{\Pi_1} = p_1\alpha_1 \overline{w}^{k_w}.\overline{\boldsymbol{X}}^{k_x} - p_w\overline{w} - p_x.\overline{\boldsymbol{X}}$$

With $\overline{w}$ and $\overline{\boldsymbol{X}}$ are input quantities without risks, and w and $\boldsymbol{X}$ are the same variables under risks with $w < \overline{w}$.

Assuming they face an exogenous risk of *water shortage*, farmers have two basic alternative options to respond to exogenous risks; to improve their water use efficiency, or to change towards alternative activities. If farmers face exogenous *water quality risks*, such as salinity or other exogenous pollution that threaten their production, they may engage into treatment, drainage or dilution, or other water pollution mitigation practices or opt to change their activities. We will therefore present the three cases separately (efficiency, alternative activity and water quality response).

When considering the first option of changing practices or technology towards efficiency (noted *e*), the profit functions expressed as follows.

$$\Pi_e = p_1\alpha_1 w_e{}^{k_{we}}.\boldsymbol{X}^{k_x} - p_w w_e - p_x.\boldsymbol{X} - C_e \quad (4)$$
$$\overline{\Pi_e} = p_1\alpha_1 \overline{w_e}^{k_{we}}.\overline{\boldsymbol{X}}^{k_x} - p_w\overline{w}_e - p_x.\overline{\boldsymbol{X}} - C_e$$

Where C_e is the fixed cost associated with the more efficient irrigation technology, assuming variable costs are not significant.

In the case of alternative activities (option noted *a*), all agronomic and cost factors change.

$$\Pi_a = p_a\alpha_a w_a{}^{k_{wa}}.\boldsymbol{X}_a{}^{k_{xa}} - p_w w_a - p_{xa}.\boldsymbol{X}_a \quad (5)$$
$$\overline{\Pi_a} = p_a\alpha_a \overline{w_a}^{k_{wa}}.\overline{\boldsymbol{X}}_a{}^{k_{xa}} - p_w\overline{w}_a - p_{xa}.\overline{\boldsymbol{X}}_a$$

In the case of water quality risk mitigation (noted *q*), assuming the same amount of water is available either way, the profit functions expressed as follows.

$$\Pi_q = p_1\alpha_1 w_q{}^{k_w}.\boldsymbol{X}^{k_x} - (p_w + c_q)w_q - p_x.\boldsymbol{X} \quad (6)$$
$$\overline{\Pi_q} = \Pi_q$$

In this case the cost c_q is assumed to be mostly variable, to represent the importance of volumes for treatment or change of practices.

Farmers will switch to more efficient practices under the following condition:

$$E(\Pi/e) \geq E(\Pi/NR) \Leftrightarrow \varphi[(\Pi_e - \overline{\Pi_e}) - (\Pi_1 - \overline{\Pi_1})] \geq \overline{\Pi_1} - \overline{\Pi_e} \quad (7)$$

$$\Leftrightarrow C_e \leq \varphi p_1 \alpha_1 \boldsymbol{X}^{k_x}({w_e}^{k_{we}} - w^{k_w}) + (1-\varphi) p_1 \alpha_1 \overline{\boldsymbol{X}}^{k_x}(\overline{w_e}^{k_{we}} - \overline{w}^{k_w}) + p_w[\varphi(w - w_e) + (1-\varphi)(\overline{w} - \overline{w}_e)] \quad (6)$$

Thus the expected cost of switching to higher efficiency technology has to be smaller than the expected net revenues it will generate. Given that the efficiency condition implies an increase of productivity under risks, i.e., ${w_e}^{k_{we}} > w^{k_w}$, but not necessarily in the absence of risks, so $\overline{w_e}^{k_{we}} \leq \overline{w}^{k_w}$, the probability that inequality (6) stands (noted θ_e) increases with the probability of water risks and the cost of water, but not necessarily with other variables (as these factors increase the right hand side expression of the inequality, assuming all other terms remain constant).

With simplification assumptions of water is de facto free for farmers, no change in water use under risks or no risks, and the same input uses, (7) simplifies to:

$$E(\Pi/e) \geq E(\Pi/NR) \Leftrightarrow K_e \leq p_1 \alpha_1 \boldsymbol{X}^{k_x}\left[\varphi({w_e}^{k_{we}} - w^{k_w}) + (1-\varphi)(\overline{w_e}^{k_{we}} - \overline{w}^{k_w})\right] \quad (8)$$

In this case, the probability θ_e will also increase with prices and input quantities, given that they increase the right hand side expression of (8).

Farmers will switch to alternative crops under the following condition:

$$E(\Pi/a) \geq E(\Pi/NR) \Leftrightarrow \varphi[(\Pi_a - \overline{\Pi_a}) - (\Pi_1 - \overline{\Pi_1})] \geq \overline{\Pi_1} - \overline{\Pi_a}$$

$$\Leftrightarrow \varphi\left[\left(p_a\alpha_a\left({w_a}^{k_{wa}}.{\boldsymbol{X}_a}^{k_{xa}} - \overline{w_a}^{k_{wa}}.{\overline{\boldsymbol{X}}_a}^{k_{xa}}\right) - p_w(w_a - \overline{w}_a) - p_{xa}.(\boldsymbol{X_a} - \overline{\boldsymbol{X}}_{\boldsymbol{a}})\right) - (p_1\alpha_1(w^{k_w}.\boldsymbol{X}^{k_x} - \overline{w}^{k_w}.\overline{\boldsymbol{X}}^{k_x}) - p_w(w - \overline{w}) - p_x.(\boldsymbol{X} - \overline{\boldsymbol{X}}))\right] \geq p_1\alpha_1\overline{w}^{k_w}.\overline{\boldsymbol{X}}^{k_x} - p_a\alpha_a\overline{w_a}^{k_{wa}}.{\overline{\boldsymbol{X}}_a}^{k_{xa}} - p_w(\overline{w} - \overline{w}_a) - p_x.\overline{\boldsymbol{X}} + p_{xa}.\overline{\boldsymbol{X}}_{\boldsymbol{a}}$$

$$\Leftrightarrow p_a\alpha_a\left(\varphi {w_a}^{k_{wa}}.{\boldsymbol{X}_a}^{k_{xa}} + (1-\varphi)\overline{w_a}^{k_{wa}}.{\overline{\boldsymbol{X}}_a}^{k_{xa}}\right) - p_1\alpha_1(\varphi w^{k_w}.\boldsymbol{X}^{k_x} + (1-\varphi)\overline{w}^{k_w}.\overline{\boldsymbol{X}}^{k_x}) - p_w(\varphi(w_a - w) + (1-\varphi)(\overline{w}_a - \overline{w})) - p_{xa}.(\varphi\boldsymbol{X_a} + (1-\varphi)\overline{\boldsymbol{X}}_{\boldsymbol{a}}) + p_x.(\varphi\boldsymbol{X} + (\boldsymbol{1}-\varphi)\overline{\boldsymbol{X}}) \geq 0 \quad (9)$$

The expression is hardly tractable, unless simplified. The probability that inequality (9) is verified will increase with the probability of risks and with the cost of water (as long as the use of water with the alternative is effectively lower than that of the original activity, and other variables constant).

Assuming the cost of water is negligible, and that the price of inputs is the same for the two alternative activities, condition (9) becomes:

$$E(\Pi/a) \geq E(\Pi/NR) \Leftrightarrow \varphi\left(p_a\alpha_a{w_a}^{k_{wa}}.{\boldsymbol{X}_a}^{k_{xa}} - p_1\alpha_1 w^{k_w}.\boldsymbol{X}^{k_x} - p_x.(\boldsymbol{X_a} - \boldsymbol{X})\right) + (1-\varphi)\left(p_a\alpha_a\overline{w_a}^{k_{wa}}.{\overline{\boldsymbol{X}}_a}^{k_{xa}} - p_1\alpha_1\overline{w}^{k_w}.\overline{\boldsymbol{X}}^{k_x} - p_x.(\overline{\boldsymbol{X}}_{\boldsymbol{a}} - \overline{\boldsymbol{X}})\right) \geq 0 \quad (10)$$

Condition (10) is consistent with a net revenue constraint; the probability that it is realised will increase with the relative market price of the alternative activity and decline with the possible increment in input costs it may be associated with (all other variables remaining constant).

Farmers will adopt water quality mitigation approaches under the following condition:

$$E(\Pi/q) \geq E(\Pi/NR) \Leftrightarrow -\varphi(\Pi_1 - \overline{\Pi_1}) \geq \overline{\Pi_1} - \overline{\Pi_q}$$
$$\Leftrightarrow c_q w_q \leq p_1\alpha_1\big((w_q{}^{k_w} - \varphi w^{k_w}).\boldsymbol{X}^{k_x} - (\overline{w}^{k_w} - \varphi\overline{w}^{k_w}).\overline{\boldsymbol{X}}^{k_x} - p_w\big((w_q - \overline{w}) - \varphi(w - \overline{w})\big) - p_x.(1-\varphi)(\boldsymbol{X} - \overline{\boldsymbol{X}}) \quad (11)$$

The total cost of treatment has to be lower than the mitigation of such risks for this condition to be satisfied.

Endogenous risks changes the problem, but not necessarily its incentives

With endogenous risks, the action of water using farmers at time t_0 affects whether there will be a risks in the future, i.e. $\varphi = f(\sum_{k=1}^{n} w_0^i)$ with w_0^i representing the use of water (or water pollution in the case of water quality) of farmer i at time t_0. Typically farmers' use—or misuse—of water will not explain all the risks, but contribute to it. One could assume for instance a linear expression of water risks:

$$\varphi = \varphi_0 + (1-\varphi_0)\left(\frac{\mu}{W_0}\sum_{i=1}^{I} w_0^k\right) \quad (12)$$

Where φ_0 represents exogenous water risks, $0 \leq \mu \leq 1$ is a parameter capturing the future impact of water use on available resources for the sector and W_0 is the maximum water usable by agriculture at t_0. This formulation implies that $0 \leq \varphi_0 \leq \varphi \leq \mu + (1-\mu)\varphi_0 \leq 1$.

Such specification will change (1) to become:

$$E(\Pi/R) \geq E(\Pi/NR) \Leftrightarrow \left[\varphi_0 + (1-\varphi_0)\left(\frac{\mu}{W_0}\sum_{i=1}^{I} w_0^i\right)\right]\left[(\Pi_R - \overline{\Pi_R}) - (\Pi_1 - \overline{\Pi_1})\right] \geq \overline{\Pi_1} - \overline{\Pi_R} \quad (13)$$

Thus farmers' water use (or pollution) at time t_0 will increase the probability of their reaction in the future, *ceteris paribus*.

A farmer's response will not necessarily change from the exogenous risk case, because the probability of future risks depends on the use (or impact) of *all* farmers at t_0. An individual farmer reducing water use or water impact earlier, by switching to activities or practices, will reduce water risks in the long term, thereby reducing the need (and probability θ) for *any* farmer to invest in future response. However, unless the farmer accounts for a significant share of water use in the hotspot region (e.g. as part of a collective of users), or that s/he faces a sufficiently strong signals to account for future water risks (policy or investment return), s/he may not be sufficient incentivised to act.

The dynamics of the problem are extremely simplified in the model; farmers in hotspot regions may already face risks, and therefore invest early, but the aggravation of risks in the future that they may cause is unlikely to trigger a sufficient incentive to react to future water risks. This is even more the case when acknowledging that multiple variables remain uncertain in the future.

4.A1.3. Modelling the company's response

There are several differences with the previous model, because companies are purchasing farmers' products, because they may affect prices of agriculture outputs, and because these products may only represent one input in their production process. The companies also do not face the same type of impact; in the case of exogenous risks, they may face future production losses, whereas for endogenous risks, they may also face reputation risks, which could reduce their sale and the confidence of investors.

Under exogenous water risks, shifting supplies or ingredients will mitigate risks at a cost

Profit functions at time t_1 are defined as follows:

$$\Pi_1^c = P_1(\sum_i y_i)^{k_y^c}.\boldsymbol{Z}^{k_z^c} - p_1(\sum_i y_i) - P_z.\boldsymbol{Z} \qquad (14)$$

$$\overline{\Pi}_1^c = P_1\left(\sum_i \overline{y_i}\right)^{k_y^c}.\boldsymbol{Z}^{k_z^c} - p_1\left(\sum_i \overline{y_i}\right) - P_z.\boldsymbol{Z}$$

With P_1 as the output price, $\boldsymbol{Z}$ as other inputs, at price P_z, and y_i as the agriculture product quantity of farmer i ($i \in [\![1, N]\!]$). Given the first stage of the model, some farmers may have decided to respond to risks by increasing efficiency or changing to other activities. It is assumed here that we focus on the N farmers that have decided to continue producing the sourcing ingredient for companies (with $N \leq I$). Of these N farmers, a share σ_e has adopted more efficient water management practices, and similarly a share σ_q has taken measures to mitigate water quality risks.[2]

Under exogenous risks, water is not a main input for their production, and water risks can therefore only affect farmers' production in the hotspot region. Three types of responses are likely: (a) a shift in products outside of the hotspot region, (b) a change of agriculture input within the same region that do not face the same risks, and (c) supporting $(1-\sigma)$ N farmers that have not invested into water efficient practices. Naturally the size of the company will likely determine whether it makes sense to use option (a) or (c), and the type of product will also matter in determining the feasibility of (b). Here we take a hypothetical case where the three possibilities exist.

Under the first option —shifting purchases to other farmers (noted *out*)— an individual company's profits can be derived as:

$$\Pi_{out}^c = P_1(\sum_{l=1}^{L} y_l)^{k_y^c}.\boldsymbol{Z}^{k_z^c} - (p_1 + c_t)(\sum_{l=1}^{L} y_l) - P_z.\boldsymbol{Z} \qquad (15)$$

$$\overline{\Pi}_{out}^c = \Pi_{out}^c$$

Assuming the same quantity can be purchased outside of the region, the companies' economic impacts of water risks are mitigated but there is a transport cost c_t for any new product.

Under the second option —changing inputs (noted *inp*) — profits are the following:

$$\Pi_{inp}^c = P_1(\sum_{m=1}^{M} v_m)^{k_v^c}.\boldsymbol{Z}^{k_z^c} - p_v(\sum_{m=1}^{M} v_m) - P_z. \qquad (16)$$

$$\overline{\Pi}_{inp}^c = P_1(\sum_{m=1}^{M} \overline{v_m})^{k_v^c}.\boldsymbol{Z}^{k_z^c} - p_v(\sum_{m=1}^{M} \overline{v_m}) - P_z.\boldsymbol{Z}$$

With v_m representing the production of farm m.

Lastly, in the third option, companies might decide to engage into efforts to reduce water use with the non-efficient farmers, proposing to bear an acceptable share ρ of the costs of investment (ensuring that the remaining farmers satisfy condition (7)). Profits are then:

$$\Pi_e^c = P_1(\sum_i y_{ei})^{k_y^c}.\boldsymbol{Z}^{k_z^c} - p_1(\sum_i y_{ei}) - P_z.\boldsymbol{Z} - \rho N(1-\sigma)K_e \qquad (17)$$

$$\overline{\Pi}_e^c = P_1\left(\sum_i \overline{y_{ei}}\right)^{k_y^c}.\boldsymbol{Z}^{k_z^c} - p_1\left(\sum_i \overline{y_{ei}}\right) - P_z.\boldsymbol{Z} - \rho N(1-\sigma)K_e$$

Where y_{ei} is the production of efficient producer i ($i \in [\![1, N]\!]$).

The arbitrage between the three options and the original one will depend on the cost of transportation, trade-off between more costly ingredients and risks of losing production, and on the cost of supporting farmers to become efficient.

- Option 1: Procuring products out of the region

$$E(\Pi^c/out) \geq E(\Pi^c/NR) \Leftrightarrow \varphi.\left[\left(\Pi^c_{out} - \overline{\Pi^c_{out}}\right) - \left(\Pi^c_1 - \overline{\Pi^c_1}\right)\right] \geq \overline{\Pi^c_1} - \overline{\Pi^c_{out}}$$

$$\Leftrightarrow P_1\left((\textstyle\sum_{l=1}^{L} y_l)^{k^c_y} - \varphi(\sum_i y_i)^{k^c_y} - (1-\varphi)(\sum_i \overline{y_i})^{k^c_y}\right).\mathbf{Z}^{k^c_z} - p_1(\sum_{l=1}^{L} y_l - \varphi \sum_i y_i - (1-\varphi)\sum_i \overline{y_i}) \geq c_t(\sum_{l=1}^{L} y_l)$$

$$\Leftrightarrow c_t \leq \frac{1}{\sum_{l=1}^{L} y_l}\left[P_1\left((\textstyle\sum_{l=1}^{L} y_l)^{k^c_y} - \varphi(\sum_i y_i)^{k^c_y} - (1-\varphi)(\sum_i \overline{y_i})^{k^c_y}\right).\mathbf{Z}^{k^c_z} - p_1(\sum_{l=1}^{L} y_l - \varphi \sum_i y_i - (1-\varphi)\sum_i \overline{y_i})\right] \tag{18}$$

The cost of transportation has to be smaller than the benefits of mitigating the risks, this condition will hold especially with high risks, high damage with risks.

- Option 2: Changing inputs (ingredients)

$$E(\Pi^c/inp) \geq E(\Pi^c/NR) \Leftrightarrow \varphi.\left[\left(\Pi^c_{inp} - \overline{\Pi^c_{inp}}\right) - \left(\Pi^c_1 - \overline{\Pi^c_1}\right)\right] \geq \overline{\Pi^c_1} - \overline{\Pi^c_{inp}}$$

$$\Leftrightarrow$$

$$P_1\left[\varphi\left((\textstyle\sum_{m=1}^{M} v_m)^{k^c_v} - (\sum_i y_i)^{k^c_y}\right) + (1-\varphi)\left((\sum_{m=1}^{M} \overline{v_m})^{k^c_v} - (\sum_i \overline{y_i})^{k^c_y}\right)\right].\mathbf{Z}^{k^c_z} - [p_v(\varphi \sum_{m=1}^{M} v_m + (1-\varphi)\sum_{m=1}^{M} \overline{v_m}) - p_1(\varphi \sum_{l=1}^{L} y_l + (1-\varphi)\sum_i \overline{y_i})] \geq 0 \tag{19}$$

The expression is hard to track, but if one assumes that the production of the alternative input is not affected with water risks it simplifies to:

$$E(\Pi^c/inp) \geq E(\Pi^c/NR) \Leftrightarrow P_1\left[(\textstyle\sum_{m=1}^{M} v_m)^{k^c_v} - \varphi(\sum_i y_i)^{k^c_y} - (1-\varphi)(\sum_i \overline{y_i})^{k^c_y}\right].\mathbf{Z}^{k^c_z} - [p_v \sum_{m=1}^{M} v_m - p_1(\varphi \sum_{l=1}^{L} y_l + (1-\varphi)\sum_i \overline{y_i})] \geq 0 \tag{20}$$

In this case, it becomes another risk mitigation strategy which will be cost-effective only if the additional cost of the ingredient is lower than risk avoidance.

- Option 3: Contracting with farmers

$$E(\Pi^c/e) \geq E(\Pi^c/NR) \Leftrightarrow \varphi.\left[\left(\Pi^c_e - \overline{\Pi^c_e}\right) - \left(\Pi^c_1 - \overline{\Pi^c_1}\right)\right] \geq \overline{\Pi^c_1} - \overline{\Pi^c_e}$$

$$\Leftrightarrow \varphi\left[P_1(\textstyle\sum_i y_{ei})^{k^c_y}.\mathbf{Z}^{k^c_z} - p_1(\sum_i y_{ei}) - P_1(\sum_i \overline{y_{ei}})^{k^c_y}.\mathbf{Z}^{k^c_z} + p_1(\sum_i \overline{y_{ei}}) - P_1(\sum_i y_i)^{k^c_y}.\mathbf{Z}^{k^c_z} + p_1(\sum_i y_i) + P_1(\sum_i \overline{y_i})^{k^c_y}.\mathbf{Z}^{k^c_z} - p_1(\sum_i \overline{y_i})\right] \geq P_1(\sum_i \overline{y_i})^{k^c_y}.\mathbf{Z}^{k^c_z} - p_1(\sum_i \overline{y_i}) - P_z.\mathbf{Z} - P_1(\sum_i \overline{y_{ei}})^{k^c_y}.\mathbf{Z}^{k^c_z} - p_1(\sum_i \overline{y_{ei}}) - P_z.\mathbf{Z} - \rho N(1-\sigma)K_e$$

$$\Leftrightarrow K_e \leq \frac{1}{\rho N(1-\sigma)}\left[P_1\left[\varphi\left((\textstyle\sum_i y_{ei})^{k^c_y} - (\sum_i y_i)^{k^c_y}\right) + (1-\varphi)\left((\sum_i \overline{y_{ei}})^{k^c_y} - (\sum_i \overline{y_i})^{k^c_y}\right)\right].\mathbf{Z}^{k^c_z} - \left[p_1\left(\varphi(\sum_i y_{ei} - \sum_i y_i) + (1-\varphi)(\sum_i \overline{y_{ei}} - \sum_i \overline{y_i})\right)\right]\right] \tag{21}$$

This condition will be more favourable if the share of costs and number of farms is smaller, if the risk for production is greater, and if water efficiency really creates effective solutions to water risks.

While this is outside the scope of the exercise, on a small market, and with a limited number of companies, the decision of one company to opt for one of these solutions or to remain in a default will affect the price of the raw commodity. In particular, if a significant share of companies decides to opt for other inputs, the price p_1 may decline, which could result in less incentives for farms to respond, and in turn affect the non-reacting companies.

Companies are further encouraged to engage with farmers under endogenous water risks

In this case, three factors change. First, farmers' decision to react may change (equation 13) and instead they take action earlier; secondly, their action will affect overall risks which could create uncertainties; and third, relatively large companies that do not react could face risks to their reputation that in turn could affect their sales (loss of revenues) and/or confidence of investors (which would affect long term growth and investments). These risks can be factored in as penalties $g(\varphi)$ depending on the risk φ:

$$E(\Pi^c/NR) = \varphi\left(\Pi_1^c - \overline{\Pi_1^c}\right) + (1-\varphi)\left(\Pi_1^c - \overline{\Pi_1^c}\right) - g(\varphi) \qquad (22)$$

With $\varphi = \varphi_0 + (1-\varphi_0)\left(\frac{\mu}{W_0}\sum_{i=1}^{I} w_0^k\right)$ and $\frac{dg}{d\varphi} \geq 0$

Everything else being equal, larger companies are therefore more likely to engage in a response than smaller companies. In practice, larger companies will also have better access to knowledge and technologies, and may be able to realise economies of scale, increasing the likelihood that they will respond.

Notes

1. The cost of water is set to be exogenous and mostly representing the cost of access. In the case of water markets these prices could evolve with conditions, which would change the outcome.

2. These two shares could be endogenous if prices of agriculture outputs were affected by the choice of companies. To keep things simple, they have been set as exogenous.

ORGANISATION FOR ECONOMIC CO-OPERATION AND DEVELOPMENT

The OECD is a unique forum where governments work together to address the economic, social and environmental challenges of globalisation. The OECD is also at the forefront of efforts to understand and to help governments respond to new developments and concerns, such as corporate governance, the information economy and the challenges of an ageing population. The Organisation provides a setting where governments can compare policy experiences, seek answers to common problems, identify good practice and work to co-ordinate domestic and international policies.

The OECD member countries are: Australia, Austria, Belgium, Canada, Chile, the Czech Republic, Denmark, Estonia, Finland, France, Germany, Greece, Hungary, Iceland, Ireland, Israel, Italy, Japan, Korea, Latvia, Luxembourg, Mexico, the Netherlands, New Zealand, Norway, Poland, Portugal, the Slovak Republic, Slovenia, Spain, Sweden, Switzerland, Turkey, the United Kingdom and the United States. The European Union takes part in the work of the OECD.

OECD Publishing disseminates widely the results of the Organisation's statistics gathering and research on economic, social and environmental issues, as well as the conventions, guidelines and standards agreed by its members.

OECD PUBLISHING, 2, rue André-Pascal, 75775 PARIS CEDEX 16
(97 2016 44 1 P) ISBN 978-92-64-27954-4 – 2017

IWA Publishing's authorised EU representative for General Product Safety Regulations is Diane D'Arras, 15 rue Duret, 75116 Paris, France, e-mail: safety@iwap.co.uk.

Printed and bound by CPI Group (UK) Ltd, Croydon, CR0 4YY

12/05/2026

02108892-0003